Halff of Texas
A Merchant Rancher
of the Old West

Patrick Dearen

EAKIN PRESS ✥ Fort Worth, Texas
www.EakinPress.com

Copyright © 2000
By Patrick Dearen
Published By Eakin Press
An Imprint of Wild Horse Media Group
P.O. Box 331779
Fort Worth, Texas 76163
1-817-344-7036
www.EakinPress.com

1 2 3 4 5 6 7 8 9
ISBN-13: 978-1-57168-436-3

Library of Congress Cataloging-in-Publication Data

Dearen, Patrick
   Halff of Texas : Merchant Rancher of the Old West / Patrick Dearen. -- 1st ed.
      p.   cm.
   Includes bibliographical references and index.
   ISBN 1-57168-436-0
   1. Halff, Mayer, 1836-1905. 2. Ranchers--Pecos River Region (N.M. and Tex.)--
Biography. 3. Ranchers--Texas--Biography. 4. Ranch life--Pecos River Region (N.M. and
Tex.)--History--19th century. 5. Frontier and pioneer life--Pecos River Region (N.M.
and Tex.) 6. Merchants--Pecos River Region (N.M. and Tex.)--Biography. 7. Ranching--
Pecos River Region (N.M. and Tex.)--History--19th century. 8. Texas--Biography. 9.
Pecos River Region (N.M. and Tex.)--Biography. I. Title
F392.P3D433   2000
976.4'061'092--dc21
   [B]   CIP
                                                                      00-059293

For James Owens,
who shares my oneness with Creation

Other Books by Patrick Dearen

Nonfiction
Castle Gap and the Pecos Frontier
Portraits of the Pecos Frontier
Crossing Rio Pecos
A Cowboy of the Pecos
The Last of the Old-Time Cowboys

Novels
The Shakwa (co-author, pseudonym Gene March)
Starflight to Faroul
The Illegal Man
When Cowboys Die

Contents

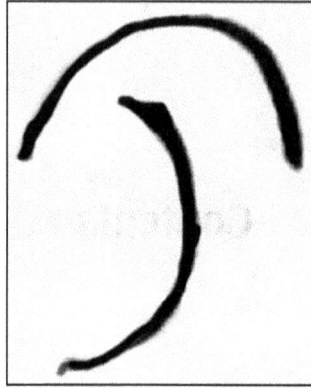

Mayer Halff's Quien Sabe brand.

The Quien Sabe Ranch as painted in 1899 by L. Schloss.

—Courtesy Witte Museum, San Antonio, Texas

Author's Note

In November 1996 I met with three of Mayer Halff's descendants, who offered their cooperation and support as I wrote a book about Mayer's life and times. For two years I researched and wrote in close consultation with Mayer's grandson, Jesse Halff Oppenheimer, and two great-grandsons, Howard A. Halff and Alex H. Halff.

My quest for facts carried me from the Big Bend and Pecos River of Texas to musty archives to newspaper microfilm in Kansas. It has been a fascinating search, for Mayer Halff was a remarkable man who hewed several of the Old West's most famous ranches out of little more than wilderness.

Quotes attributed to Mayer regarding his ranching activities are as related by Quien Sabe cowhand Bob Beverly, who knew Mayer personally. Quotes attributed to Mayer regarding his home life are as handed down through family tradition.

Mayer Halff (1836–1905)
—Courtesy Alex H. Halff

Chapter 1

Out of France

For sixty-nine years, cowman Mayer Halff rode a long, busy trail that few men could have followed.

It carried him from 1850s Europe to Texas and across fifteen hundred miles of wild western country waiting to be tamed. Throughout, he was at ease, whether dining with a U.S. president at a plush metropolitan banquet, or squatting across a campfire from a dirt-streaked cowboy in some forsaken cow camp. He was a respected man who in turn respected others, a man driven to succeed and possessed of the character and business acumen to do so.

Mayer's deeply scored trace began along the Lauter River in Lauterbourg, France, where he was born on February 7, 1836, to Jewish parents, Henry Halff III and his wife, Eve. Situated in the province of Alsace, the village of his birth lay quite near the German border. The valleys there were green and inviting, the mountains undemanding, and dotting the agricultural plain were almost two hundred villages, eleven of them within a twenty-nine-mile radius of Lauterbourg.

For almost a millennium, the region had been home to a Jewish minority. Barred from living in the larger French cities until the late 1700s, many Jews continued to choose the simple country life in Alsace in the 1830s. Like most Alsatians of Jewish descent, Mayer's father had roots in the province extending

1

back generations, and Judaism's rich heritage there was a buffer against persecution.

More than Judaism was in young Mayer's blood, however, for he was the son, grandson, and great-grandson of livestock traders. Through tradition and experience, Jews had long ruled the Alsatian cattle industry, from breeding and pasturing to butchering. Religious ritual dictated the method of slaughter, to ensure Jewish use of the meat, but the demand by non-Jews was nonetheless great. As soon as the sun set on every Sabbath and ended sanctions against trading, Jewish butchers regularly opened their doors to other peoples anxious for beef.

The cattle industry undoubtedly dominated Mayer's boyhood. Morning after morning, he may have opened the gates of his father's cattle sheds and, with the lure of a shepherd's pipe, led the animals to the common summer pastures. On occasion, he may have accompanied his father to the well-known cattle markets in St. Die and Lure. Cattle, as well as money lending, had elevated his parents to middle or upper-middle class, a mark of relative affluence for Lauterbourg Jews, and in the hoofs, horns, and hides Mayer may have already seen his life's vocation.

Still, there were other traditions almost as influential. Two of Mayer's uncles were Lauterbourg merchants, and his great-grandfather had been a peddler—a trade still vital in Jewish society. With their goods slung over their backs, peddlers traveled the countryside afoot, stopping at villages and farms to hawk needles, handkerchiefs, almanacs, and other items manufactured in the cities. In a day without mass communication, they also brought news from afar to the *Landjede,* or country Jews.

Although Mayer learned to read and write Hebrew, a necessity for a Jewish boy's synagogue studies, he likely spoke a Jewish-German dialect of French with his parents, his older sisters Babette and Rosalie, and his younger brother Solomon. Solomon, born May 8, 1838, seemed to learn new languages easily, an attribute that would prove invaluable as he pursued a common destiny with Mayer in America.

In 1849 Mayer turned thirteen, an important milestone in Jewish life. Until then he had merely studied the law and the 613 mitzvoth, or commandments, in order to understand the obligations of being Jewish. In a Sabbath morning ceremony, he would be charged with the adult duty of observing these rules. The bar mitzvah brought a Jewish boy before the synagogue congregation for the first time, either to chant the Torah lesson or to read from the haftarah, or Prophets. Following the ritual, the rabbi marked him as a "son of the commandment." Not only did it entitle him to don the tephillin, or phylacteries, for prayer each morning, but in the eyes of the Jewish community he was a man, with all the religious responsibilities of such.

In young Mayer's first year or so of religious majority, circumstances beyond his control set the stage for his future. Economic depression raged throughout Europe, crippling opportunities for a boy suddenly turned man. There was political uncertainty as well, for France and Germany both claimed Alsace, and war was not beyond possibility. Whether a conflict erupted or not, conscription in the French army of Louis Napoleon Bonaparte was almost a certainty for any Alsatian teenager.

Already, Mayer's first cousin Adolphe Halff, a Lauterbourg native about the same age as Mayer, had immigrated to North America and settled in Texas. Even in its days as a republic, Texas had been well known to the French, who had recognized its sovereignty and had maintained an embassy in its capital. Texas was now part of the United States, a country with assured religious freedom and incomparable opportunity. From a base in Galveston, the most important port between New Orleans and Veracruz, Mexico, Adolphe had initiated a successful peddling business. As events would demonstrate, Adolphe or his family back in Lauterbourg must have told Mayer of the opportunities in Texas and urged him to join his cousin in the venture.

And so it was that in about 1850 fourteen-year-old Mayer, a boy by today's standards but a man in nineteenth-century Jewish society, said goodbye to the only home and family he had ever known, boarded a ship, and set sail for a strange new world. Even at so tender an age, he was displaying the vision, initiative, and courage that would drive him for the next half-century.

When he reached Galveston, an island in the Gulf of Mexico two miles off the Texas coast, he found a bustling town (with a small Jewish population) where a merchant or drummer could readily obtain merchandise imported from industrialized regions. In the entire hemisphere, there was probably only one familiar face—Adolphe's—but Mayer's sense of adventure likely assuaged his loneliness. Soon, he and his cousin were both peddling, probably in a joint operation.

Many decades later, peddling client Lucinda Winfree (later Dunman) remembered Adolphe in particular; Mayer's life on the road likely paralleled his cousin's. Traveling a circuit through towns and plantations, Adolphe frequently showed up at the Winfree door, fifteen miles from Liberty, with goods he had brought by pack animal from Galveston. Often, her family opened their home to him for the night, for he was well liked, especially by the plantation owners of French descent who appreciated his mother tongue.

For several years, Mayer pursued a similar lifestyle, assimilating the customs and language of a land far removed from Lauterbourg. At times, perhaps in communities, he no doubt stuffed his merchandise in a pack and went door-to-door on foot, just as his Alsatian forbears had done for centuries. Although no documentary evidence exists that he ever owned a slave, a credible report survives that at this time he enlisted a Negro as an assistant.

"Old William," as Mayer called this man whose last name may have been Sheppard, was destined to play an important role in Mayer's eventual life on the range. For now, however, any notions of entering the cattle business were submerged in Mayer's burgeoning merchandise business. Already, he and Adolphe may have secured a storage building at Liberty Landing on the Trinity River fifty miles north of Galveston. In 1856 they made plans to make the jump from drummers to resident merchants.

The nearby town of Liberty, situated between the Gulf prairies and the dark, forbidding woods of the Big Thicket, had much to offer young entrepreneurs. Up and down the Trinity, slaves worked the fertile fields of large plantations owned by wealthy Creoles, while the forests were opening up before the saws of lumbering operations. Although twenty miles inland, Liberty stood at the head of the navigable portion of the Trinity and was an important port, connecting the region's enterprises with the world's shipping lanes. Regularly, steamers freighted cotton, tobacco, sugar cane, lumber, and hides to Galveston and returned bearing mail, passengers, and any wholesale items a merchant might need. Furthermore, Liberty sat astride the Old Spanish Trail, a busy route to and from Nacogdoches, the Louisiana border, and New Orleans. Even then, Liberty buzzed with talk of possible rail service, and within two years the Texas and New Orleans Railroad would be a reality.

Probably no city in Texas held more promise for new merchants, and Mayer's road to financial success would run right through its heart.

By mid-spring 1856 Mayer and Adolphe were ready to do business inside an existing Liberty store run by a Judge Branch. The cousins announced the opening in an April 28 newspaper ad that offered readers insight into their business philosophy: "Our motto being 'Cheap for Cash,' we will not be undersold by any store in the place."

Although the advertised stock indicated a concentration on dry goods, Mayer already was displaying a smart businessman's instincts for diversifying: The ad closed with a call for hides, deerskins, and beeswax, for which the cousins would pay "the highest market price."

Brisk business on through midsummer encouraged the cousins to expand their stock, and at 5:00 P.M., August 8, Adolphe set sail from Galveston on the steamer *Nautilus* to purchase goods in northern (or perhaps European) markets. Also on board were twenty-nine other passengers, one hundred horses, seventy cattle, and $30,000 in coin.

That night, despite a lack of wind, the tide rose to an unusual level on Galveston Island, suggesting a hurricane in the Gulf of Mexico. When the *Nautilus* failed to arrive on schedule at New Orleans, authorities feared the worst. Soon the steamer *Perseverance* embarked from New Orleans for Galveston; searching for clues, the crew found the coast from Sabine Pass to Point Bolivar strewn with wreckage and livestock carcasses.

The *Nautilus* had gone down, carrying Adolphe to his death. Only a man buoyed by a cotton bale survived.

The blow to Mayer must have been devastating. He had lost not only a countryman, a cousin, and a friend, but also the business expertise of someone who had grown up the son of a Lauterbourg merchant. Likely, Mayer had heretofore deferred to

Adolphe's experience in many facets of their store's operation; now every detail was on Mayer's shoulders alone, and the suddenness, shock, and grief must have redoubled his professional burden.

Coincidentally, however, another Alsatian relative with experience and initiative was suddenly on the scene. From Mayer's perspective, the parallels between the newcomer, twenty-six-year-old Felix Halff, and Adolphe were astounding. Felix, too, was a first cousin and a native of Lauterbourg, where he had grown up in the home of another of Mayer's merchant uncles. In 1851 Felix had immigrated for reasons familiar to Mayer: To avoid conscription and to seek opportunity. For a year and a half, Felix had peddled in Louisiana; then he had entered the merchandising business in St. Landry's Parish with Alexander Blum. After eighteen months, Felix and Blum had dissolved their partnership, but Felix had continued the enterprise until closing out sometime in 1856. Perhaps spurred by the death of Adolphe, he relocated to Liberty and went into business with Mayer.

The nature or duration of their partnership is unclear. The 1860 census found Felix in Liberty, but with only $100 in property. At any rate, it was yet another Halff from Lauterbourg whose fate would be entwined with Mayer's in the mercantile business.

As early as Adolphe's death, Mayer may have penned a letter to his brother Solomon and encouraged him to immigrate. By this time, Solomon was eighteen, refined, and noted for his kindness. He was also intellectual, having attended some of the finest schools in Europe in pursuit of a teaching career. Moreover, he was a master linguist, with eventual proficiency in French, Hebrew, German, English, and Spanish. He would be an asset to Mayer in more ways than one.

While Solomon mulled over his brother's invitation, Mayer expanded his Liberty operation to include groceries. In *Liberty Gazette* ads of December 1856, he boasted "a fine stock . . . of staple and fancy goods" and solicited hen eggs at fifty cents per dozen.

On February 7, 1857, Mayer observed his twenty-first birthday. For years already he had considered himself a man. He had journeyed to a strange country, fended for himself, and established a successful business. Now he was an adult physically and legally in the eyes of American society. Standing five feet, six inches, he was fair skinned and had dark eyes and a shock of wavy brown hair. He may have already sported the brown beard he would display in his late twenties. Furthermore, he had begun to exhibit the character traits that would mark his later life.

The only extant photograph featuring Mayer as the sole subject is notable for a smile that seems natural, unforced, as if he were accustomed to wearing it. Indeed, his contemporaries described him as a genial man who made many friends and remained loyal to them. Even as he attained wealth and stature, he retained his earthy demeanor and never forgot his relatively modest beginnings. On the contrary, he displayed an altruistic side and often gave financial help to the needy in Lauterbourg or to worthy causes in America.

With his employees, he was approachable, friendly, and ready with a helping hand. "He had a great deal of feeling for his men," said Bob Beverly, one of his cowboys.

The December 24, 1905, *San Antonio Daily Express* described this employer-employee relationship eloquently: "The humblest Negro porter in the big wholesale establishment and the wildest cowboy on his ranches each felt that he had in Mr. Halff a friend who would listen sympathetically and deal justly."

In all phases of life, Mayer was respected for his honesty and his unsullied name. He seemed above even the hint of scandal, though he did number among his friends men whose own reputations were less than spotless. Although a man of generally happy disposition, he was never content to rest on the fruits of past success. Indeed, he seemed compelled to take an active role, driven to new heights by his restless edge.

Like any man, however, he had traits that sometimes led those around him to lose patience. For one thing, his ranch hands didn't always know what to make of his frugality. For a dollar-a-day cowboy to be thrifty was one thing; for a rancher of enormous wealth to wander the range picking up scrap metal or soup bones was something else.

Furthermore, letters to Mayer from his son Henry (in 1903) and from Solomon (in 1867) suggest that he sometimes wanted to control a situation, even though he had delegated responsibility to someone else. Perhaps it was simply the psyche of a self-made man, this occasional reluctance to trust even his own son or brother to handle a matter with as much experience and expertise as he could. And his expertise was considerable.

"He was many years ahead of his day and time," recalled Beverly. "[He] taught his men to save and accumulate [and] practiced what he preached. . . . Mr. Halff gave me good advice, and I still love him for it."

In 1857 Solomon abandoned his plan to teach and decided to join Mayer in Liberty. Booking passage on a ship bound for North America, he reached the United States sometime after May 8, his nineteenth birthday. Upon his arrival in Liberty, he almost certainly shared a residence with Mayer, a situation that would continue at least into the spring of 1860. However, Solomon's role in Mayer's Liberty operation was not as a partner but as a bookkeeper.

Mayer, his morale no doubt boosted by Solomon's presence, set about in the summer of 1857 to expand his business even more. In the July 13 *Liberty Gazette*, he advertised the recent arrival of "a large and superior stock" of dry goods, perfume, jewelry, and groceries. Again, he assured patrons of his competitive values ("I am determined to sell . . . at Galveston prices"), but he also agreed in print for the first time to accept "cotton or hides" in lieu of money.

This concession in a region rich in production but poor in cash was significant, for it was the first step to accepting cattle as payment—a development that eventually would lead to Mayer's ranching empire.

Now that Mayer was of legal age, he was obligated to pay county and state taxes. For 1857 he declared five horses, two wagons, and merchandise worth $2,500, a lofty sum for the era. Although it is true that he had assumed Adolph's share of the business, it is nevertheless clear that he had made a rapid and successful transition from drummer to merchant.

By 1858 Mayer had acquired for $150 a city lot that was almost certainly the location for his mercantile establishment. Whether this was the old Judge Branch store, in which he and Adolphe had opened their business, or a subsequent structure is not clear. However, a brick building with roots traceable to

Mayer still stood at the corner of Trinity and Main streets in the late 1990s. It included a high floor (to facilitate easy unloading by wagon freighters) and a fireplace in the back wall. By 1859 Mayer opened a branch store in Sumpter, seventy miles north-northwest of Liberty.

Little is known of Mayer's personal life in the late 1850s, but his advertisements in the *Liberty Gazette* suggest that his business flourished. In a long-running ad first placed on April 1, 1859, he introduced new items—saddlery, hardware, and cutlery—and continued to promote his previous line of goods. "In fact," it read, "Everything usually kept in a store may be found at MEYER [*sic*] HALFF'S."

The year 1860 proved pivotal for twenty-four-year-old Mayer, although it started out routinely enough. He had made the acquaintance of William B. Duncan, a Liberty County cattleman who traded at his Liberty store. Mayer borrowed $50 from Duncan on February 16 and repaid him four days later. Duncan in turn squared up his account at Mayer's store with a payment of $4.85 on March 18. In the not-too-distant future, the two men would cross paths again.

In the meantime, however, Mayer observed one of the most important events of his life. A subject of France ever since he had immigrated a decade before, he now sought naturalization as an American. On May 3, 1860, he presented the required evidence to Liberty County Court, renounced his allegiance to France, and took the oath of citizenship. Ironically, he would hold citizenship less than a year before his adopted state would secede from the Union.

An indication of Mayer's swelling affluence came on June 6, 1860, when the U.S. census taker recorded his household. Solomon, yet to claim an interest in his brother's business, was included in the record only as a bookkeeper. Mayer, however, stood out for his financial worth—$19,000, of which $5,000 was in real estate. The fact that for 1860 Liberty County valued his town lot at only $2,500 suggests that he owned real estate of equal value somewhere else. Considering the developments of the next several months, it is possible that he already had acquired pasturage for cattle.

Assuredly, Mayer had come to a decision that summer of

1860. Despite his success as a merchant, he evidently longed to leave the confining atmosphere of a store and start a new life under the open sky. For a long while already, he may have accepted cattle in exchange for merchandise, and any such herd demanded attention. But more importantly, in the bellowing of cattle he may have heard a summons that fired his imagination in a way the haberdashery world never could.

The Liberty region was already rich in cattle tradition. Twenty or so years before, James Taylor White had founded a ranch at Turtle Bayou in then-Liberty County (now Chambers County) and had declared 1,775 cattle for tax purposes. Later, he had driven herds to New Orleans. Encouraged by his success, other ranchers had helped boost Liberty County's cattle numbers from 14,058 in 1840 to 45,670 in 1850.

The most daring step on Mayer's road to becoming a cowman came in the summer of 1860, when he sold his mercantile business to Alphonse Dreyfus, who was married to Felix Halff's sister Julie. Thus, through marriage, thirty-one-year-old Julie became Mayer's fourth Lauterbourg relative to play a role in his Liberty operation.

Dreyfus and Julie had immigrated through New Orleans on November 14, 1856. They had lived in Arnaudville and Grand Coteau in Louisiana before relocating to Liberty by June 1860, when the U.S. census was taken. Interestingly, the census taker found the couple sharing a home with Felix.

Announcing the takeover of the Halff store in an ad placed in the August 31, 1860, *Liberty Gazette,* Dreyfus noted that he would "continue the business at the old stand of said Halff." Whether Mayer retained ownership of the premises is not known.

What is certain, however, is that the way was now clear for Mayer to pursue what would be his passion—the mossy horns and flinty hooves of an animal ready to stampede hellbent-for-leather through a romantic era in American history.

Chapter 2

From Hats to Hooves

By the late 1850s, Liberty County cattlemen were driving thousands of cattle annually to New Orleans, the first significant market for Texas beef. The unforgiving trail, popularized in the early 1840s, harbored mosquitos, ticks, and biting flies, and carried herds through swampy land and across rivers and bayous. Small shuck fields, available for a price, generally provided the only grazing, a situation that limited the size of herds to only a few hundred head. Along the route, settlers manned way stations where drovers could lodge overnight and pen their herds for a nickel a head.

When a trail boss finally pointed his herd into New Orleans, he exchanged the animals for gold and rode for Texas. The beeves, meanwhile, were either targeted for slaughter, sold to buyers in outlying regions, or shipped to the West Indies.

East Texas cattlemen also used steamship travel for delivering herds to New Orleans, although rates were expensive and a beef lost weight on the journey. Once Texas seceded from the Union on March 2, 1861, however, federal ships blockaded gulf ports. The abrupt end to sea shipments sent an unprecedented rush of herds down the trail for New Orleans. Mayer rode at the point of one such drove.

May 29 found Mayer's cattle penned at Day's way station, about a day's drive from Liberty. The size of his herd is not known, but two trailing droves numbered 240 and 375 head, respectively. Having gone from peddler to merchant, Mayer was now a cowboy, bossing a modest trail outfit. A typical Louisiana-bound herd required but three drovers and a very small remuda, although this latter fact often forced a cowhand to ride the same mount from dawn till dusk. When hunger called in this pre–chuck wagon era, drovers looked to pack horses trailing with the cattle.

By June 30 Mayer and his cowhands had driven to New Iberia, Louisiana. There, Liberty resident William Duncan, boss of a trailing herd, caught up with Mayer and encouraged him to hold his drove so that they could proceed together. Nevertheless, Mayer chose to push on. At Lyon's Point, fifty miles farther downtrail and thirteen days from Liberty, Duncan again met up with Mayer. Evidently, the site marked the terminus of Mayer's drive, as he was preparing for the ride back to Liberty. On July 1 Duncan presented him with a letter to carry back to Duncan's wife in their hometown.

Further evidence of Mayer's cattle interests in this initial year of the Civil War appears in the 1861 Liberty County tax roll. For the first time, Mayer declared cattle in the county, paying taxes on fifty head valued at five dollars each and on twelve horses worth an aggregate of $300.

The 1861 Liberty tax roll is also notable for the first inclusion of Solomon, who now owned 264 acres valued at $1,600. In all likelihood, Mayer's livestock grazed his brother's land.

Although Mayer no longer operated a business in Liberty, he may have retained a stock of goods. He, or perhaps his former store, sold Duncan eighty-two pounds of flour for $5.53 on October 22 and another $5.50 in flour on December 4. Mayer also engaged in money lending, a trade that had helped make his father relatively affluent in Alsace. On November 6 he loaned Duncan, who was preparing for another drive, $200 on a thirty-day note. On December 5 Duncan dutifully repaid him.

February 1862 evidently found Mayer in Louisiana, for he wrote Ed White of Liberty a letter that prompted White and Duncan to consider embarking for Louisiana on business. Several weeks later on March 26, Duncan, who kept a careful diary, last made reference to Mayer. Bound for Houston, Duncan waited at the "Half [*sic*] store" for the train to arrive. Duncan, out of habit, may have been referring to the new Dreyfus operation by its former name.

The year 1862 brought another pivotal moment for Mayer, who, in face of war, chose to relocate. On August 21 Liberty County officials included him on resident tax rolls for the last time. Solomon, meanwhile, soon found himself caught up in military affairs of which he evidently wanted no part. On December 29 he and Alphonse Dreyfus were drafted by the Confederate States of America. Neither had been naturalized; legally, they remained French citizens. When they failed to report as ordered, a Confederate detail arrested the pair and placed them under guard. On January 15, however, they appeared before P. L. Palmer, district clerk in Liberty County, and filed a statement claiming exemption from the draft.

"[We] are both subjects of the Emperor Napoleon," their document stated, "and therefore claim French protection."

By asserting their allegiance to France, the two men successfully avoided military service. However, Solomon's rightful claim to French citizenship over a matter of conscription would eventually come back to haunt him.

Mayer, no longer a Liberty resident by 1863, was nevertheless included in the Liberty County tax roll for that year. Officials levied taxes on two tracts of land he had acquired in nearby counties: 1,995 acres in Jefferson County and 25 acres in Harris County.

Mayer's whereabouts and activities for the remainder of the Civil War are largely a mystery, although he, like Solomon, did not serve in the military. It is known that Felix Halff engaged in cotton trade in Brownsville and in the Mexican cities of Matamoras and Piedras Negras. Mayer also apparently entered the cotton business, trading in South Texas and northern Mexico. In November 1863, as Union troops approached a small garrison of Confederates at Fort Brown in Brownsville, the Confederates torched the fort and its cotton stores before retreating. Evidently, 3,800 pounds of the cotton belonged to Mayer, for a November 3 Confederate document indicates that Mayer was due that amount.

With Union troops occupying Fort Brown, Mayer (and perhaps Solomon) ventured into Mexico, where he may have continued dealing in cotton. A resident of Matamoras by early 1865, Mayer secured a passport from the local French consulate on March 13 for passage to France. It is unclear if business played a role in his decision to travel abroad, but he evidently had not seen his family in Alsace since his departure a decade and a half before.

The Civil War ended in 1865, but whether Mayer remained in Matamoras or returned to Texas immediately is not certain. Solomon, however, took up residence in San Antonio in time to be listed in the 1865 Bexar County tax roll.

Late in the year, Mayer resurfaced in Texas with ambitious plans. San Antonio had become the dominant city in the state. Moreover, it was gaining a reputation as the hub of the Southwestern cattle industry. Even during Reconstruction, the town offered the kind of opportunity that Liberty had promised a decade before. Perhaps influenced by Solomon, Mayer moved to San Antonio and established a dry goods business with his brother and their relative by marriage, Abraham Levy. Levy was the husband of Mayer's first cousin Esther, who was the sister of Felix Halff.

The first entry in a Halff & Levy ledger is dated November 25, 1865, but the trio may have organized their operation several months earlier. An 1865 Texas brand book, *The Texas Stock Directory*, includes an advertisement for "Halff & Levy, Wholesale and Retail Dealers in Fancy and Staple Dry Goods," located on Commerce Street in San Antonio. Advertised items included clothing, hats, boots, and shoes. The ad also noted that Halff & Levy would pay the "highest cash prices . . . for hides."

The new partners first advertised in the *San Antonio Herald* on December 19. An accompanying news article noted their "superior stock of goods . . . [offered] at reduced prices." Early patrons of their store included Roy Bean, who would gain notoriety in Langtry as Judge Roy Bean. On January 9, 1866, Bean purchased a pair of pants for $2.51, a shirt for $3.00, an overshirt for $3.50, drawers for $4.00, socks for fifty cents, and a handkerchief for twenty-five cents.

The Halff & Levy enterprise was an immediate success, and the three men went to lengths to establish themselves. In early

1866 Solomon ventured to New Jersey and New York, evidently on business, and stayed for an extended period. By January 1869 Halff & Levy would be worth more than $100,000 in merchandise and profit shares, a giant step from Mayer's days at the Liberty store. In the meantime, though, Mayer took an even greater leap in his personal life.

In February 1866 he observed his thirtieth birthday. He already had accomplished a great deal. Still, he was alone, even in the company of his brother and Levy. But in New Orleans— a city that Mayer may have frequented on business—waited a young Jewish woman named Rachel Hart.

Born October 10, 1845, in New Orleans, Rachel had eight siblings, including a fraternal twin. Described by those who knew her as charming, sweet, and popular, Rachel lived a life "full of generous thoughts and noble deeds," according to her Rabbi. Her parents, Isaac Hart and the former Julia Cohen, had emigrated from England before marrying. Sometime in 1866, Rachel and her family moved to Detroit, Michigan, far removed from affairs in San Antonio. Nevertheless, her future soon would be entwined with Mayer's.

Whether Mayer met Rachel during her years in New Orleans is not known. More likely, a Jewish matchmaker was involved, for Mayer was only a few years removed from Europe, where most Jewish marriages were arranged. The tradition was so strong among Alsatian Jews that it often accompanied them to the New World. The matchmaker, whether in Europe or the United States, diligently observed strict class lines in setting up marriages; in return for his services, he received a fixed percentage of the required dowry.

Regardless of the circumstances of their engagement, on September 2, 1866, Mayer and Rachel were married before a Rabbi in Detroit and Mayer soon took his bride to live in San Antonio.

In the spring of 1867, Solomon visited his family in France. Evidently, French military orders calling for his conscription caught up with him during his stay. He had successfully argued for his exemption from the Confederate draft on the basis of his French citizenship; now that very allegiance found him in a similar dilemma across the Atlantic. He wrote Mayer of his situation, informing him that "the price of a substitute is now 3,000 francs. . . . If I can procure me the necessary funds, I might go and give myself up." With the firm of Halff & Levy so successful, it is something of a mystery why Solomon would have had difficulty acquiring the requisite money.

Although the details are unclear, Solomon somehow escaped conscription and returned to the United States.

Important milestones occurred in Mayer's personal life in the late 1860s and 1870s. He and Rachel became parents on September 25, 1867, when their daughter Henrietta, or Hennie, was born. Five months later, Mayer purchased three San Antonio lots on Goliad Street. At what would become 139 Goliad, they would build a house in 1893 and make their home there for the rest of their married lives.

Late 1868 found Rachel expecting their second child, and for the final period of her pregnancy she returned to her parents' home in Detroit. There on February 5, 1869, a son, Alexander Hart, was born. Another son, Henry Mayer, who would be instrumental in Mayer's livestock business, was born August 17, 1874; and on August 4, 1877, Rachel gave birth to their daughter Lillie. (Two other children, Rosa and Sidney, would die in infancy.) Mayer, it seemed, had settled down—yet his initiative and restless spirit would carry him far afield in succeeding years.

In the spring of 1870, Solomon once more visited Alsace—and in so doing again found himself facing conscription. On May 17 he was ordered to report for duty at Strasbourg. Nevertheless, in circumstances that are vague, he evidently succeeded once more in avoiding the draft. Three weeks later, with his return to America imminent, he was elected to the board of directors of a San Antonio insurance company, Western Texas Life, Fire, and Marine.

Personal milestones were also taking place in Solomon's life: On November 15, 1871, he married Fannie Levi of Victoria, Texas, who was described as a kind and affectionate woman. They became parents of a son, also named Henry, in 1873, and a daughter, Minnie, in 1875. A third child, Mayer Leo, was born in 1878, followed by Godchaux A. C. Halff in 1879 and Cecile two years after. Solomon would also build a home on Goliad Street, almost directly across from Mayer's. The two houses, incorporated into Hemisfair 1968, remained preserved in Hemisfair Park at the turn of the twenty-first century.

In the 1870s Mayer and Solomon experienced major business developments. In 1872 they dissolved their partnership with Levy, advertising a closeout sale in the *Freie Presse für Texas*, a German-language newspaper published in San Antonio. Reason for the sale, as noted in the ad, was "an intended change in our previous business."

But Mayer's role in the dry goods industry was still in its formative stages. Immediately, he and Solomon organized M. Halff & Brother, which was destined to become a leading haberdashery and wholesale operation in the Southwest. The declared value of their initial stock, as stated on the 1873 Bexar County tax roll, was $20,000. By the summer of 1875, the brothers had opened an office in New York, and within another three years, the value of their San Antonio headquarters had more than tripled to $62,450. By 1890, Bexar officials appraised their San Antonio holdings at more than $104,000, an impressive amount for the day.

With growth came a need for better facilities in San Antonio. In January 1877 M. Halff & Brother moved into a new building on Commerce Street adjacent to the Schmitz Hotel. On August 28, the *Freie Presse für Texas* described the finished structure as "the prettiest, most pleasant, and most elegant of its type . . . in San Antonio. . . . It is an architectural master work." The article went on to say that the establishment was one of the largest in the region and was virtually "without equal in the state. . . . The move of the firm to their new facility will begin a new era for the firm of M. Halff & Bro."

From humble beginnings as a peddler, Mayer had risen to a dominant position in the wholesale and dry goods world. Now, his pocketbook swelling, he was primed to hew himself a niche of even greater importance in the livestock industry.

Chapter 3

On the Trail to Dodge

Money was scarce in Texas immediately after the Civil War, but cattle were not. Millions of longhorns ran wild in the brush, for Texas was possibly the best breeding country in a reunited nation filled with beef-hungry people. With an inexhaustible demand on the one hand and a self-perpetuating reservoir of beef on the other, only one problem remained: transportation.

Interregional railroads were yet to link Texas directly with markets such as St. Louis or Chicago, but in the hooves of the longhorn lay the means of bridging the gap between the South Texas brush country and the railroads of Kansas. Too, the Great Plains of the north waited untrampled, the tall grass prairies offering ideal ranges for maturing cattle. In the late 1860s and 1870s, Kansas became the grazing and shipping headquarters for Texas stockmen, who annually drove thousands of cattle to cow towns such as Abilene, Salina, Baxter Springs, Coffeyville, Ellsworth, and Dodge City.

The latter town first gained prominence as a terminus on the Western Trail out of Texas in 1876. Situated in the Arkansas River Valley in the southwestern part of Kansas, Dodge City offered cattlemen shipping opportunities via the Atchison, Topeka & Santa Fe Railroad. By 1878 Dodge City boasted a population of 1,000 and had assumed several sobriquets, all revolving around the cattle industry. Cattlemen knew it as the "best mar-

18

ket in Kansas," the "cattle market of the West," "the great Texas cattle market of America," and the "go-ahead little city of the West." Cowboys who sought to unwind a little after several months on a drive knew it better as "the liveliest town in Kansas." Indeed, for several weeks every summer when Texas cowhands crowded the saloons, few places in the West were as wild. Fisticuffs in particular were so commonplace that they often went unnoticed. "Street fights too numerous to mention for the past week," noted a Dodge City newspaper in July 1878.

Not all Texas cattlemen chose to ship their beeves immediately upon reaching Dodge City. The surrounding prairie offered excellent maturing grounds, and Dodge City served as a logical springboard for trail bosses who decided to push their herds farther north to Ogallala, Nebraska, and elsewhere.

In 1877 the flinty hooves of approximately 200,000 cattle cut a trace up from Texas to Dodge. Some of the droves, likely composed of several respective sub-herds, were immense. The July 13 *Waco Daily Examiner* reported that of the 100,000-plus cattle in the immediate vicinity of Dodge City, 40,000 were in a single herd, with 21,000 in a second and 17,000 in a third.

Among the drives that marched up the trail to Dodge in 1877 was one of unknown size owned by Mayer, who probably had purchased the cattle specifically for the drive. Although he was now forty-one, old for a drover, he evidently accompanied the animals.

Managing a herd demanded not only insight into cattle psychology, but planning, anticipation, and the ability to act quickly. Upon throwing together a typical herd of 2,500 head, a trail boss carefully positioned his drovers. Two experienced hands rode "point" on either side of the lead and kept the animals on course. At rear, three cowboys rode "drag," prodding the slower cattle onward, while "swing" men came up along each side of the main body. Normally, cowhands "graze-drove" the animals, allowing them to forage as they drifted downtrail.

For weeks on end, drovers virtually lived in the saddle. Even sundown brought little relief, for cowhands had to ride guard in shifts around a herd often itching to stampede.

"When night came, we just penned our cows by holding our feet in the stirrups," recalled Bob Beverly, one of Mayer's drovers in the 1890s. "Whether it rained or snowed, it was all the pen we had for the cattle. The night horse had a good deal to do with whether we lived or not."

Upon reaching Dodge with his herd in 1877, Mayer sold it to a ready buyer. The *Ford County Globe,* which was published in Dodge, reported that Mayer was "well pleased" with the sale and that he planned a return drive in 1878, for "Dodge is good enough for [him]."

By March 5, 1878, Mayer had accumulated another 8,000 Texas cattle in partnership with a man named Bishop (possibly Rufe Bishop) and was ready to point the animals north. These steers and breeding stock had plenty of company, for another 257,000 Texas cattle were poised to join them in their march to Dodge City and beyond.

The massive drive required the services of 1,300 drovers, who, along with 250 owners, would flood Dodge until their business was completed. By the end of May, the leading herds raised a dust over the town, and within another month, Mayer had arrived with the Halff-Bishop drove. The partners sold 4,000 head to E. B. Millett, who himself had come up from Texas with 9,000 beeves. The fate of the other 4,000 Halff-Bishop cattle is not known.

Mayer did not start back for Texas immediately, instead traveling by train to Denver on apparent cattle business. On September 11, R. J. Lauderdale met up with Mayer in what the Texas cowhand described as a gambling house that sported faro, monte, keno, and wheel of fortune. The next morning, Mayer and Lauderdale boarded an eastbound train en route to San Antonio. The first night out of Denver, they encountered an agitated crowd of more than a hundred people at Buffalo Rail Station; only hours before, warmongering Indians had crossed the track nearby on a bloody ride north.

By a circuitous route, Mayer and Lauderdale reached a section house near Houston, where a conductor requested that all passengers display their health certificates. Yellow fever was decimating Memphis and New Orleans, and Houston authorities were taking no chances. When Mayer and his fellow travellers failed to produce the documents, the conductor put them off the train. Finally, with the assistance of a Houston doctor and telegraph lines, they established that the train had not stopped in any fever-plagued area. The next day Mayer continued toward San Antonio.

Mayer and Bishop dissolved their partnership by early 1879. With Bishop now teaming with R. G. Head for the coming drive, Mayer turned to Solomon. By March 2, the brothers had acquired 3,000 cattle to add to the estimated 197,000 head, mostly young Gulf Coast steers contracted to Dodge buyers, which Texas outfits would road-brand for the trail.

Mayer's drive of '79, which probably got underway two to three weeks late due to dry conditions and thin cattle, was notable for the first known involvement of R. A. "Rufe" Moore of McMullen County, Texas, in his affairs. About this time, Moore also had connections with a Rio Grande ranch, the Two Moon or Quien Sabe, which would later play a prominent role in Mayer's life. Already, Mayer may have seen in Moore not only a good cowman, but a close friend to whom he could entrust his cattle business. Asa Jones, who once worked for Moore, described him as "a typical old South Texas ranchman . . . , stout, good-humored, knew his cattle, horses and cowboys and knew the cow works in and out." Although Mayer's ranching empire was only in its formative stages in 1879, Moore was already the best candidate to serve as his eventual general manager.

Even now, Moore was in a position of responsibility. Although the first herds to arrive in Dodge City in June brought higher prices than the year before, Mayer had other plans for his drove of 3,000. Upon gaining the cow town, Mayer purchased another 2,000 head from Volley Oden, who had also driven up from Texas. A man named Joe Collins bought into the Oden herd as well, but whether he did so in partnership with Mayer or separately is not clear.

After R. J. Lauderdale tallied Mayer's purchase, Moore took charge, turning the beeves in with Mayer's original drove and proceeding north for Nebraska.

By July 8, 108,700 cattle and 1,500 horses had reached Ogallala on the Union Pacific Railroad, prompting a *Sidney* (Nebraska) *Telegraph* correspondent to remark, "Cattle till you can't rest." But the stampede for Ogallala was only beginning, for another 100,000 cattle were poised to enter the cow town any day. Ogallala, observed the *Telegraph* that summer, was "the net in which a few ten thousand head have been corralled."

Like Dodge City, Ogallala lived and died by the hooves of Texas cattle. Situated on the South Platte River near the point where Nebraska wrapped around northeastern Colorado, Ogallala was "the busiest burg in the west" and "the liveliest town of its size on the road," said a Sidney newspaper of the era. Other writers were less kind, one of them proclaiming Ogallala "the Gomorrah of the cattle trail." And for good reason.

"Dancing, Spanish monte, and other 'side dishes' are served up night or day," observed the *Sidney Telegraph* on July 26, 1879. "The streets are lined with saddled and unsaddled ponies, sold or for sale; the merchants, hotel men, saloon men, and every other man are busy as bees in May. . . . It is almost impossible to get even . . . lodging on the floor, and as for beds there isn't a spare one . . . in the town."

The Platte River bottom was wide, making for excellent places to hold herds until they could be delivered on contract, sold, or shipped east. In 1879 alone, more than 2,200 cars of cattle rolled out of Ogallala en route to Chicago. Thousands of other cattle went to Indian agencies or to ranches in Nebraska, Colorado, Wyoming, the Dakotas, and Montana. With two million dollars in summer cattle business, Ogallala indeed lived up to its reputation as the "Cowboy Capital."

In July 1879 the little town of thirty buildings incorporated and instituted ordinances to maintain order under the influx of hundreds of entertainment-starved cowboys from Texas. Law officers required cowhands to relinquish firearms upon entering the city. The move was a necessity, for "the ends of the earth's iniquity . . . gathered in Ogallala," wrote Andy M. Adams in *The Log of a Cowboy*.

It was upon such a scene that Mayer descended with his herd of 5,000 beeves in late July 1879. His cattle, observed the *Sidney Telegraph*, were young and of excellent class. The ultimate disposition of his drove is not known.

As Mayer prepared for the 1880 drive, he found it necessary to offer higher prices for cattle than in years past. In mid-February 1880, Victoria-area stockmen were demanding $6 a head for yearlings, $10 to $12 for two-year-olds, and $17 to $18 for three-year-olds. Three-fourths of the cattle were bound for Dodge City, despite concerns that the payout in Dodge and other cow towns would not justify higher costs in Texas.

By March, Texas stockmen had targeted 300,000 cattle for the drive, including 100,000 contracted to buyers in locations such as Dodge City and Ogallala. Good rains had opened the possibility of an early drive, for the cattle had wintered well and grazing along the trail was plentiful. Although a subsequent winter storm hampered plans for an early start, by March 12 Mayer and Solomon had readied 3,000 to 6,000 beeves for the push north.

Mayer also entered into partnership with R. G. Head for the coming drive, and their cowboys soon burned an "apple" road brand—a circle with a stem protruding to the right—on thousands of cattle. By early spring, a Halff & Head drove of 1,900 head was on the march for Mason County. The herd passed through the town of Mason on April 24 and soon reached Hedwig's Hill, a German settlement five miles south of Art in southern Mason County. There, a Mrs. Martin, likely Anna Henriette Mebus Martin, purchased the cattle, which the *News* of Mason described as "in better condition than any we have before noticed this spring."

Meanwhile, another Halff & Head drove, consisting of 2,250 to 2,500 mixed cattle, forged steadily northward. Boss of the outfit was Giles Fenner.

The first Texas herds of 1880 reached Dodge City in early May, several weeks ahead of the Halff & Head drove. By the week of May 27 to June 2, when Mayer's partnership herd marched through Fort Worth, 131,953 cattle had already passed the Texas city, double the number by the same date in 1879.

By the latter part of June 1880, an astounding 175,000 cattle and horses had reached Dodge City. So had Mayer, either in concert with one of his herds or in advance, for June 22 found him registered at a Dodge hotel.

Within another three days, the Halff brothers' herd, now numbering 3,300, and the Halff & Head drove, now totalling 2,200, were among the 205,311 cattle which had ambled into Dodge City that season. But Mayer's role in the drive of '80 was not yet over, for he and Head had targeted their herd for other points in Kansas and Nebraska. Meanwhile, the Halff drove was bound for Ogallala.

Possibly, Mayer purchased additional cattle in Dodge for the push on to Nebraska, or else he and Head divided their partnership herd. At any rate, the *Sidney Telegraph* reported on July 17 that 5,000 Halff and Brother cattle were now on the trail. Whether Mayer had contracted with a buyer or intended to mature the animals on an Ogallala-area range is not known.

Significantly, however, the *Kansas City Price Current* soon reported a shortage of grass in the South Platte Valley near Ogallala. One problem was that potential ranges had shrunk by twenty-five per cent, while cattle numbers had increased by fifty percent. "There is no use blinding ourselves," said the *Price Current*, "to the fact that the line of civilization is steadily drawing around the Texas cattle trade."

Although the maturing ranges of northern regions were in no danger of disappearing, the years of the cattle drive were rapidly drawing to a close. Soon homesteaders would crowd the trails, and intercontinental railroads would join Texas with markets such as Chicago or St. Louis. Perhaps it was time for a visionary such as Mayer to consider the possibilities inherent in vast breeding grounds of his own.

Chapter 4

Ranching the Big Bend

Out west of the Pecos River in Texas sprawled a majestic and forbidding country known as the Big Bend. Here, where the Rio Grande cut a great arc through the Chihuahuan Desert, an environment both harsh and enchanting supported cacti and untrampled grasses. It was still an untamed wilderness in the 1870s, ruled by rattlers, Indian raiders, and outlaws. Nevertheless, farsighted cattlemen began to take notice, for the range was vast and ready for the taking.

Mayer's interest in the Big Bend dated from as early as 1873, a time when cavalry troops from Forts Davis, Stockton, and Concho were still seeking to control the Comanches. W. T. Marshall and freighter John D. Burgess had already acquired a tract in Presidio County, and Mayer was interested in it on behalf of M. Halff & Brother. Throughout their long partnership, Mayer and Solomon seemed to have observed a division of duties: Solomon concentrated on the dry goods business and Mayer ruled the cattle operation.

On December 19, 1873, the brothers acquired the Marshall-Burgess property by trustee deed. It consisted of approximately 1,560 acres, the root tract of a far-flung Big Bend ranching enterprise which would grow to gigantic proportions in the 1880s. The acreage likely offered exactly what Mayer was seeking—a water source from which to reign over a broad swath of country. Water was scarce in the Chihuahuan Desert, and to control a spring or flowing creek was to rule the arid land alongside. By 1889, the brothers' Presidio County holdings totalled 4,411 acres valued at $9,535.

Their agent for tax and perhaps managerial purposes was William Russell, but the nature of the operation is unclear. From 1877 (the year of the first extant Presidio County tax roll) on through 1900, neither the Halffs nor Russell declared livestock in the county. The venture nevertheless may have concerned cattle, as tax officials sometimes overlooked even great numbers of stock.

Some distance to the east, Mayer established one of the Chihuahuan Desert's truly impressive ranches—the Peña Colorado or, by brand, the Circle Dot. His involvement in the spread dated to March 1882, when M. Halff & Brother purchased a tract from M. F. Corbett, along with 1,280 acres for $2,000 from Mr. and Mrs. Thomas K. Harnsberger of Rockingham County, Virginia. The latter property lay along isolated Maravillas Creek, forty-seven miles southeast of Fort Davis.

Originating in the Del Norte Mountains southwest of present-day Marathon, Maravillas flowed southeast through crags and desert flats, passing the 6,521-foot Santiago Peak on a long, tortuous course to the Rio Grande. With a portion of Maravillas now under his control, Mayer was in a position to dominate a wide stretch of arid country.

His next step was to secure possession of the chief spring in the area. Not far from the point where Maravillas slithered out of the Del Nortes, it met the waters of Peña Colorado Creek. Ten miles upstream along the latter waterway, a gap squeezed through imposing bluffs; immediately beyond this pass in the Marathon Basin, a spring burst from the base of a color-banded cliff 4,220 feet above sea level.

Early U.S. Army reports knew this rock face, which loomed up four miles south-southeast of present-day Marathon, as Rainbow Cliff, but it became better known as Peña Colorado, Spanish for "red rock." The cliff lent its name not only to the spring, but to the south-trending creek and eventually to Mayer's ranch.

As the most reliable water source between Comanche Springs at Fort Stockton and the Rio Grande, Peña Colorado had been a noted landmark on the Comanche War Trail. By the light of the "Comanche moon," war parties had splashed across Horsehead Crossing, skirted Comanche Springs, and passed Peña Colorado en route for Mexico. But the Comanche threat was over, and Peña Colorado stood as a potential oasis for cattle by the thousands.

Mayer was not alone in recognizing the location's value; before him, stockmen, farmers, and the U.S. Army had all ventured to Peña Colorado. The spring was situated on so-called Section 400, fifty-five miles southeast of Fort Davis; the first owner of record was a U.S. Army lieutenant named Davis. When military authorities proposed to establish a Fort Davis sub-post at the site in 1879, the lieutenant agreed to a rent-free arrangement.

On August 27, 1879, Company F of the Twenty-Fifth Infantry marched in from Fort Stockton, seventy-five miles to the northeast, and set up stakes at what would be known as Camp Peña Colorado. Company B, Tenth Cavalry, and Company G, Twenty-Fifth Infantry, soon followed, and by 1880 the soldiers occupied several mud-and-stone huts in the hill-defined bowl immediately north of the mountain gap. Roofed with mud and grass, the crude structures required repairs after every significant rain. Enlisted men lodged in a narrow sixty-three-foot hut, while officers dwelled in two single-room buildings. Additional structures included two storehouses, a granary, and a picket corral.

Although Peña Colorado was a near-paradise in many ways, it was denigrated by some observers. Burr G. Duvall, passing through in March 1880, described the military reservation as "a most desolate looking place, [with] not a tree or shrub in sight and barren hills all around."

Nevertheless, cattlemen such as E. M. Beckwith, who had once ranched on the Pecos, realized Peña Colorado's promise, especially now that it harbored hungry soldiers. On June 30, 1881, Beckwith contracted with the army to supply beef to the camp for eight cents a pound.

Sometime after the sub-post was established, Lieutenant Davis relinquished his claim on Section 400 to Torres Irrigation and Manufacturing Company. In January 1882 Francis Rooney acquired the property, but he continued to let the army occupy the location without charge. The following August, however, M. B. "Nub" Pulliam secured legal rights and demanded that the army withdraw.

Before authorities took action, Mayer and Solomon stepped in and purchased Section 400 from Pulliam for $7,500. Almost simultaneously, the brothers acquired 32,000 acres on the Pecos River one hundred miles to the northeast; Mayer obviously was not a man who dreamed or acted small. In reporting the purchases of the two "first-class" ranches, the *Texas Live Stock Journal* of December 23, 1882, noted:

"We are glad to see friend Meyer [*sic*] settle himself permanently and become a cattle raiser instead of trader. Your cows and land will make you richer quicker than the dry goods business, friend Halff, and you can sit back in your easy chair and laugh at the merchants who started when you did."

Section 400's value was no secret to county tax officials, who appraised the 640 acres at $6,000, or almost $10 an acre—ten times the rate of other tracts in the area. Nevertheless, it was money well spent, for Peña Colorado offered the three essentials to a successful cattle ranch: grass, water, and protection.

Mayer soon moved to stock the Circle Dot with South Texas cattle. In March 1883 he went to McMullen County, south of San Antonio, to oversee the transfer of 2,500 young animals to Peña Colorado, almost certainly by rail. During the same month, Mayer expressed his thoughts on a hotly debated topic— was rail service supplanting the drive as a means for getting cattle to northern markets? In announcing that he would send no herds by trail in 1883, he gave readers of the *Texas Live Stock Journal* a glimpse into the mind of not only a student of the cattle industry, but a man of vision.

"Mr. Halff," said the *Journal*, "thinks . . . that shipments by rail will be the means used in future years to get stock to market. Those who want stock will come . . . [to Texas], thus virtually placing a market at every pasture gate." The article added

that Mayer perceived only a single drawback to rail transport—
"exorbitant freight rates."

By June of 1884, Mayer had been proven prophetic, for at
least 25 percent of Texas cattle movements that spring had been
by rail. Furthermore, the Colorado *Live Stock Record* reported
that events of 1884 had demonstrated that "it is cheaper to ride
than to walk." Indeed, what had been a three-month journey by
arduous trail to Colorado had been reduced to a ride of less
than a dozen days.

"The young Broadhorn that had been rounded up in the
Panhandle of Texas on a Monday morning," said the *Record*, "[is
now] cropping the grass of Colorado on the Saturday following."

Even stalwarts such as George W. Littlefield, who had driven herds north from Texas for a decade, now agreed with
Mayer. "I am tired of the trail," said Littlefield in 1884. "This
year will be my last. In the future I will ship by rail. The annoyance of crossing the Indian reservation and getting through the
settled-up portions of the country with our cattle is enough to
drive a man crazy."

However, rail transport did involve drawbacks other than
cost. On the trail, herds usually were "graze-driven"—allowed to
forage as they marched. Over a trek of seven hundred miles, a
beef might actually gain weight. Freighted cattle, however, tend-
ed to shrink an average of one hundred pounds per steer
between Texas and Chicago. Moreover, by indiscriminately
crowding the young cattle with the old, the strong with the weak,
rail officials ensured that losses would result. Nevertheless, such
drawbacks were outweighed by rail's rapid transit, the absence of
tariffs imposed by reservation Indians, and the reduced danger
of spreading Texas fever to range cattle. Even as early as the fall
of 1884, the *Ford County Globe* of Dodge City bemoaned the
imminent end of the trail drive era. "Going up the trail in the
spring," said the *Globe*, "has been a looked-for job by the cowboys, and one which has afforded them a change from every
ranch."

Mayer was never afraid of change, however, and he contin-
ued to approach his cattle enterprise with a businessman's eye.
On May 8, 1883, he reached an agreement with Major J. G. C.
Lee, quartermaster, to lease Section 400 to the U.S. Army and

allow Camp Peña Colorado to stay afloat. Mayer was to receive $50 a month throughout the lease, which was subject to renewal every July 1 beginning in 1884. Moreover, he reserved the right to water his cattle at the springs, subject to health regulations stipulated by the post commander. Additionally, he could graze cattle and erect buildings on any portion of the reservation not required by the army.

Mayer probably realized that having a military post on the Circle Dot could be lucrative in more ways than one. Immediately, he secured two contracts to supply beef to the camp, the first for twelve-and-a-half cents a pound and the second for $7.40 a head.

In that same year of 1883, Mayer obtained patents on five nearby tracts from the State of Texas. Now, the Circle Dot, or Peña Colorado Ranch, was on its way to epic dimensions. Only a year later, the outfit would graze 2,100 cattle worth $25,200 and 30 horses valued at $750, and by 1885 the animals would range across forty-four deeded sections appraised at $24,640.

But deeded land along Peña Colorado and Maravillas creeks composed only part of the ranch, for Mayer also leased 75,000 acres on either side. With the Chihuahuan Desert less than lush, such acreage was necessary if Mayer was to maintain a large herd, which he continued to do even in droughty years. In thirsty 1886, for example, he declared 1,800 cattle (worth $16,200) and twenty-five horses (valued at $800) for tax purposes. With cattlemen traditionally declaring only a small percentage of their actual cattle, Mayer's herd was probably always larger than tax rolls indicated. Indeed, Jim B. Wilson, who settled in the region in 1883, claimed that 25,000 head once grazed the Circle Dot.

On the east bank of Peña Colorado Creek immediately south of the pass, Mayer constructed an adobe building to serve as ranch headquarters. The location, in addition to providing for water, offered two major advantages: military protection and access to a transcontinental railroad. The former was a valid consideration, for Big Bend ranchers were still under threat by Indians from Mexico, which was only forty-five miles away. In late fall 1884, in fact, a raid in Presidio County, which then included the Circle Dot, cost ranchers Wilson and Reid numerous horses.

No less vital was rail service. Only three miles north of Peña Colorado spring, the Southern Pacific track stretched east to west, and its completion in January 1883 linked the Circle Dot not only with Mayer's home in San Antonio but with distant markets. Undoubtedly, the new railroad was a major reason for Mayer's initial decision to ranch this area.

A prime example of the benefits afforded by rail transit came late that spring. With Solomon on a three-month trip to the Northeast to purchase goods, Mayer stayed in San Antonio to supervise their wholesale operation. Nevertheless, he shipped 275 cattle, ages three and up, to Chicago, where livestock agent D. C. Paxton and Company brokered a sale for $4.90 per hundred weight. Mayer believed it to be the highest price ever paid in Chicago for grass-fed cattle out of Texas.

With Circle Dot headquarters so far from the range on Maravillas Creek, cowhands soon built a rock line camp in Rock House Gap, nine and a half miles southwest of Peña Colorado. Situated immediately downstream of the Maravillas–Dugout creek confluence, the site offered excellent water and quick access to the ranch's lower country.

Heavy rains fell throughout the Big Bend in late summer 1883, and with Solomon finally back in San Antonio, Mayer had an opportunity to inspect the benefits to the Circle Dot range. But the primary reason for his visit was likely the general fall roundup, which got underway September 20 on the nearby Kincaid Ranch. This was still open country (not until 1886 would a drift fence cross the range east-to-west north of Maravillas Creek), and cowhands worked several adjacent pastures: the Halff, the Kincaid, the Gage, the Beckwith, and the Dolan. Bossing the roundup was Elkanah M. "Old Man" Herreford, manager of the Circle Dot.

Mayer soon returned to San Antonio, bearing "broad smiles and big rain stories," according to the *Texas Live Stock Journal*, but he traveled again to Peña Colorado as the roundup came to a close. He found that he had 3,000 fine, fat steers to market, and with rail service in his back yard, he looked straight to Chicago rather than to Dodge City or Ogallala. Always one to asume an active role, Mayer evidently accompanied the cattle to the Windy City in early November. On his way home, he stopped off in Kansas City on additional business, but he reached San Antonio in time to join Solomon at a December meeting that condemned fence cutting.

That fall, the *Texas Live Stock Journal* reflected on Mayer's burgeoning success. After reviewing his cattle ventures in Louisiana during his Liberty days, the *Journal* reported that Mayer now had "probably a half million in the business and seems disposed to go deeper." Nor did anyone begrudge him his success, said the periodical, for he had "the good will of all" who knew him.

For the first time in at least two years, Mayer dispatched a herd north by trail in the spring of 1884. His drove of 5,000, which probably got an early start due to favorable weather, was under contract. The identity of the buyer and the terms of their agreement are not known, but prices for contracted steers in 1884 ranged from $15.00 to $16.50 a head for one-year-olds to $23.50 to $25.00 for three-year-olds. Although Mayer preferred rail service, financial considerations probably led to his decision to drive. According to the March 18 *Ford County Globe*, many Texas cattlemen considered current shipping rates to Wichita Falls and points north "unreasonable."

In mid-July, Mayer's herd, now reduced to 3,000 and bossed by Ike Hill, passed through San Animas, Colorado, and crossed the Arkansas River. Its destination, reported the *San Animas Leader*, was Montana.

By deluging northern territories with thousands of cattle over the years, ranchers like Mayer were instrumental in making the livestock business the West's leading industry. "What ten years ago was nothing but a barren waste," observed the *Ford County Globe* on February 5, 1884, "is today producing the major portion of the beef consumed in the United States."

Not only would Mayer and Solomon send another herd by trail to Montana in 1885, according to Halff cowhand Jim Cline, but at some point they also secured an interest in a Wyoming ranch. Undoubtedly, the territory had much to offer. Its lush sea of buffalo grass, rippling in the spring wind, was cured into natural hay under the bright summer sun and provided superb maturing grounds.

John H. "Cap" White, who conducted research for Solomon's son G. A. C. Halff in the early 1940s, reported that the Wyoming outfit was Laramie Cattle Company, but few details about the enterprise are known. An 1885 brand book, published by Wyoming Stock Growers' Association, suggests a pair of candidates, however. Laramie River Cattle Company, the successor to the Whipple and May outfit, maintained headquarters in Cheyenne and pastures in Laramie County. Its cowboys burned more than a dozen brands into the hides of cattle ranging on the Laramie, Cottonwood, Chugwater, Eagle's Nest, and Deer Creek drainages. I. C. Whipple served as company president, H. M. Rogers as vice president, and H. G. Hay as secretary-treasurer.

Meanwhile, Big Laramie Land, Cattle, and Improvement Company had headquarters in Laramie City. August Trabing was general manager of the outfit, which ran cattle on the Big Laramie River and Sand Creek in Albany County. The company branded a lamp chimney under a half circle over the left hip.

Whether Mayer was associated with either ranch is uncertain, but he did seem to maintain at least a modest Wyoming presence over the years. At least two Halff droves trailed to the territory, with cowboys turning one herd loose seventy-five miles north of Cheyenne in 1886. Two years later, the *Texas Live Stock Journal* reported the late-October sale of eighty-four Halff cattle, described as "Wyoming-Texas" products, at the U.S. Yards in Chicago.

Mayer also reportedly made use of range land just below the Canadian border. Joe White, a black wrangler who hired on at Peña Colorado as a sixteen-year-old in 1882, recalled going north with ten Halff herds, at least one of them reaching Dakota Territory. Like all Circle Dot hands, White knew that Mayer was not a passive boss, but he was nevertheless stunned one day to see Mayer drive into their North Dakota cow camp in a buckboard.

In late May of 1884, Mayer made a special trip to the Circle Dot to receive yearlings he had purchased from Lon Millet. A month later, when the Circle Dot's beef contract with Camp Peña Colorado expired, ranch manager E. M. Herreford negotiated a new agreement on Mayer's behalf at eleven cents a pound. It brought Mayer steady income, for the enlisted men and officers consumed approximately fifty pounds of beef per day, or 1,572 pounds a month. In addition, Herreford contracted to provide the post with 430,000 pounds of Circle Dot hay at $16 a ton.

In view of the sweeping nature of M. Halff & Brother's cattle business, Mayer found it essential to delegate responsibility to trusted managers such as Herreford. Despite his vigor and passion, Mayer was just one man, and now he had also burned his brand deep in the bank of a river notorious for its demands —the Pecos.

Chapter 5

In the Coils of the Pecos

Down from New Mexico it coursed, a sinuous and sinister moat slashing across an arid empire burned by the sun. It had been known by many names down through the centuries—Pa-co-as Te-hu-as, Rio de las Vacas, Salado, Puerco. The river had also moved pioneers to curse it with a vengeance, for the Pecos, as it was known by the 1870s, was only one bend away from hell, and some claimed it wasn't even that far. Buffalo hunters were of the opinion that when a bad man died, he went to one of two places—hell or the Pecos.

The distinction wasn't clear because the Pecos had crushed so many lives that even iron-willed Charles Goodnight, who knew the river intimately, damned it as "the graveyard of the cowman's hopes."

The reasons were myriad. Its waters were briny and alkaline, the "world's worst," according to a 1932 study. Only occasionally was it potable, and even then it had remarkable purgative properties that sent many a cowboy scurrying for the bushes. Prone to flooding—sometimes reaching a mile wide—the river would recede and invariably leave deadly pools glistening in the desert.

Moreover, the Pecos currents were swift (like a war horse, said one traveler) and ominous in their silence. The banks were high and sheer, and an animal might wander alongside for half

35

a day without gaining access to the water. Indeed, in the 240 miles of the Texas stretch were barely a half-dozen places where a horseman or wagoner might cross. Below the unyielding banks, meanwhile, waited a thick layer of quicksand, treacherous in its tenacity and ever ready to claim cow, horse, or rider.

For buffalo hunters, even a hell had a heaven on the other side, but bordering the Pecos east and west was a land even harsher than the river. Not a single tree marked the banks, and the flood plain held only salt grass and other undesirables. Some plants, in fact, derived sustenance from ground so impure that horses could get "alkalied" in only a few minutes of grazing. Furthermore, this was Indian and outlaw country; Comanches kept the war trail to Mexico open through 1874, and white settlers still lived in fear of Apache raiders in the early 1880s.

But beyond the flood plain, amid scrub mesquite and creosote, grew nutritious grasses that lured many cattlemen to the region in the 1870s and 1880s. Mayer was one of them, and the collective courage of these ranchers forced the Pecos to yield its reputation as strictly an evil wasteland. Within two or three decades, many stockmen could join A. S. Gage in proclaiming the Pecos "the heart and center of the best breeding country in the world."

The stretch of river that Mayer chose was rugged in the extreme and lay along the Crockett-Pecos county line. The flats of Horsehead Crossing, sixty miles upstream of the present U.S. 290 bridge, gave way there to country rent by canyons. Sheer faces of rock and castellated hills towered seven hundred feet, guarding the valleys of the Pecos and south-trending Live Oak Creek, which emptied into the Pecos near modern Sheffield. Live Oak's pleasant waters were especially coveted, for all the way upriver to the New Mexico line was scarcely another decent east-side spring, much less a perennially flowing tributary.

Near the Pecos–Live Oak confluence lay old Fort Lancaster, abandoned and in ruins by the late 1870s. Nevertheless, the U.S. Army and other travelers still frequented the adjacent Lower Road, which connected San Antonio with El Paso and other points west. Some old-timers still knew the route as the California Cattle Trail, over which thousands of beeves had trekked en route to mining camps before the Civil War.

Mayer may have had a presence on the east side of the Pecos in the Live Oak Creek area as early as 1879. The *Texas Live Stock Journal* reported on December 23, 1882, that he and Solomon had "occupied" the Pecos north of Live Oak "for some time." The term *occupied* is significant, for Mayer never owned land on the Pecos prior to 1882, and no record of a lease is known. More likely, he had simply searched for and found an unclaimed stretch of river and turned his cattle loose, relying on line riders to control their drifting.

Such a cavalier attitude toward ownership was the order of the day on the Pecos. Even John Simpson, who with his partners burned the Hash Knife into the largest herd under one brand in the nation, did the same upon trailing a herd to the river in 1879. Although the Hash Knife soon controlled 105 miles of river north of Horsehead Crossing, its legal rights were nil; the outfit never owned or leased a single acre.

Mayer's initial venture on the Pecos may have been in conjunction with H. C. Tarde (or Tardy) of Del Rio, with whom he operated two ranches. One ranch was evidently adjacent to the Pecos–Live Oak confluence, near which Burr G. Duvall and his geological expedition noted Tarde's free-ranging cattle on January 16, 1880. In the fall of 1882, Halff and Tarde purchased 1,200 cattle from Francis Rooney of neighboring Pecos County for $10 a head. Mayer's talent for anticipating trends in the cattle industry was already evident; within eighteen months, the price of beeves had doubled.

Mayer's vision for the Pecos was still in its infancy, however. He realized that free grazing would soon be a thing of the past, and that the river could be claimed only by the man who established legal rights. By November 1881 he had already acquired two 160-acre tracts that he later would include in his list of Pecos lands. Then, in the fall of 1882, he dissolved his partnership with Tarde and divided their ranches. Mayer quickly moved to gain legal control of miles of east-side riverfront north of the Pecos–Live Oak junction.

But even as he negotiated with New York and Texas Land Company—the owner of record—the Pecos displayed its hellcat nature. On December 12 Dick Robertson rode into Pecos City, adjacent to the Hash Knife range, and reported that Mescalero

Apaches had stolen nine horses from the D. J. Crouch outfit and five horses from the Eddy brothers. Even in the absence of raids, said Robertson, cattlemen faced a discouraging situation.

"The water or grass, it is hard to say which, is killing cattle off by hundreds," he reported. "One man tells of seeing 35 at one watering place. During the last heavy storm, for which this country is noted, the cattle mixed very badly."

Nevertheless, Mayer and Solomon reached a December agreement with New York and Texas Land Company for the purchase of 32,000 acres along the Pecos in Crockett County for $80,000. On December 14, within days of the agreement, the brothers also acquired Peña Colorado from Nub Pulliam. Interestingly, Pulliam also closed a trade that month, acquiring 32,000 acres of Pecos County land that included twenty-nine miles of river frontage directly across from the Halff spread. Pulliam planned to move 7,000 cattle to his new ranch, which absorbed the entire range of Hart Mussey and Jesse H. Presnall, and a portion of John Strickland's.

M. Halff & Brother's deed for the Pecos acreage was not dated until February 20, 1883, when they made a down payment of $16,000. Additional payments of equal value were due at yearly intervals, the last in 1887.

Mayer's new range, which he would call the Pecos Ranch and others would dub the JM for its primary brand, took in fifty sections of land beginning immediately above the mouth of Live Oak Creek and extending twenty-five miles up the Pecos River. Each section included a half-mile of river front and extended two miles back into bottom land or hills.

Mayer, while making history of his own, had chosen a range steeped in history. The lower stretch incorporated Lancaster Crossing, one of the few fords on the Pecos and a gateway to the west on the Lower Road or California Cattle Trail. The upper boundary lay just upstream of Camp Melvin and Pontoon Bridge, ten miles northwest of modern Iraan. The bridge, put in place by the military in 1870 to serve the Upper Road, had been the river's most important crossing point until the completion of the Texas and Pacific Railroad far upstream in 1881. A small garrison at Camp Melvin had guarded the bridge from its inception, but now the camp and a nearby stage stand were abandoned and became a part of the unfenced JM.

Now that Mayer controlled a significant stretch of the Pecos, it was easy to lease the arid land on its perimeter from railroad companies and the State of Texas. Eventually he leased a reported 250,000 to 300,000 acres, including additional river frontage upstream of Camp Melvin. The terms of Mayer's early lease agreements, if any existed, are not known, but Pecos cattlemen generally paid three cents per acre a year.

Whether Mayer leased or not, the grazing was there for the taking in this open kingdom. A longhorn steer, free of worries about ownership lines, might range back twenty-five miles from the Pecos and traipse in only twice a week to drink. Beyond that distance, however, the dearth of water was a natural boundary, negating any reason for a cowboy to hold cattle in check. On the riverside, meanwhile, the troublesome banks, quicksand, and swift currents formed a barrier better than any barbed wire. Only upstream and downstream did JM cowhands have to ride line, and any JM cattle that slipped past could be reclaimed at the next general roundup.

Although cowhands established a lower headquarters across from present-day Sheffield, Mayer chose the upper end of his deeded range for his primary headquarters. The reason was probably the presence of the Upper Road, with its connections to points west (via Pontoon Bridge or a downstream ford) and east (via Grierson's Spring or Five-Mile Draw). Although Camp Melvin offered shelter, cowhands shunned the abandoned structures in favor of the old stage stand a few hundred yards away, remembered cowhand W. H. Holmsley, who hired on with Nub Pulliam across the river in 1882.

The JM and Circle Dot ranges demanded livestock, and during the same month that Mayer purchased Peña Colorado and bargained for the Pecos Ranch, he contracted for most of the available young steers in McMullen County. Cost was $12 a head for the one-year-olds and $15 a head for the two-year-olds.

Mayer was no stranger to cattle trade in McMullen, the home of his eventual general manager, Rufe Moore. The previous April, a time of heavy rains in the South Texas county, Moore had delivered one thousand one- and two-year-old cattle to Mayer for $10 to $13.50 a head. In November, perhaps in anticipation of delivery, Mayer had registered the Circle Dot and Circle S as road brands in McMullen; he would follow up by registering the JM (which was burned into an animal's left side) as a road brand on February 15, 1883. The following month, he ventured to McMullen to oversee the transfer of 2,500 contracted cattle to Peña Colorado. By 1887 Mayer and Solomon would range 700 cattle and 50 horses in McMullen, and in another four years they would own more than twenty-six sections, 1,500 cattle, and 100 horses in the county.

While some Texas cattlemen were preparing to send herds north in the spring of 1883, Mayer had other notions. Receiving the last of his contracted cattle in McMullen County on April 2, he readied to dispatch the drove to the JM to hold until the next season. The *Texas Live Stock Journal*, reporting that two other cowmen also had decided to hold rather than drive, considered it an "inaugural departure" and wise decision. "The growth of the cattle," said the *Journal*, "will give them a handsome profit on their investment."

As the JM Ranch began to swell with cattle, Mayer realized that the Pecos range was self-limiting. Without water away from the river, cattle that were young or weak or that had calves could never seek the grasses in the mountains or side canyons. By necessity, they would be forced to hug the river, where forage would be poorest due to overgrazing and undesirable plants. Mayer's solution, dating to perhaps as early as 1883, was to bring water to distant grazing lands by means of a novelty for the era: the windmill.

When Samuel David McWhorter, a carpenter by trade, ventured to the Pecos in 1883, he made his living by building ranch structures. One day the JM foreman approached him.

"I've got a windmill coming in just any day now," the foreman announced. "I'd like to get you to take it down to the river and put it over a hand-dug well we got down there."

"What's a windmill?" asked McWhorter. "I've never heard of one."

"Well," said the JM boss, "it'll be in here in a few days and you can study it over and see if you can figure how to put it up."

Soon the windmill arrived and McWhorter familiarized

himself with its operation and the requirements of putting it into place. He obtained four six-by-six-inch beams, each twenty-four feet long, and set to work building a tower at the site. Upon its completion, he had difficulty spreading the massive corner posts over the well shaft, which was four feet in diameter. Nevertheless, the enterprise proved successful, and even when McWhorter left the area sometime later, the mill was still in working order.

By the turn of the twentieth century, the fans of fifty-one JM windmills whirred in the Pecos wind, expanding the effective grazing range of every JM bovine to the outer boundary of the constantly expanding ranch. In addition, four large earthen reservoirs dotted the landscape. Each watering site had a fifty-foot cypress trough with unslaked lime and pebbly salt from Juan Cordona Lake, an important mineral bed up the Pecos in present Crane County. Eventually, the JM would store a load of salt (15,000 pounds was an average haul by burro train) in a rock structure and resupply the troughs as needed. It was especially important that cattle have a salt lick in early spring, for it helped them shed their winter hair. Moreover, with ample minerals at their disposal, cattle were less inclined to chew on bones that could lodge in their mouths and result in death by starvation.

It's not known whether Mayer sent a JM emissary to the stockmen's convention in Pecos City on March 8, 1883, but most of the prominent outfits on the Pecos were represented, including the Tarde, Dawson, C. B. Eddy, Continental, and Seven Rivers. They arranged a joint spring roundup and established a large reward for the capture of rustlers on the upper Pecos. In response to a proposal to disarm their cowboys, the cattlemen wasted no time voting it down as "inexpedient and unsafe for the present."

Spring brought excellent rains to the Pecos, blanketing the wide valleys with the finest blue gramma. The season also brought notice of Mayer in the *Texas Live Stock Journal*, which reported on May 26 that he, Nub Pulliam, and John Simpson were bringing large herds to the region because they controlled great stretches of the Pecos. "Water is hard to get," observed the *Journal*, "and those who located lands a few years ago and got the water, hold the fort."

Mayer's fort was extensive, populated by an army of cattle. The tax roll that summer for giant Tom Green County, which looked to the Pecos River for its western boundary, included a listing for 1,500 Halff cattle valued at $15,000, along with two 160-acre tracts. The inclusion of Mayer on the Tom Green roll suggests that he occupied the Pecos on the upstream in present Crane County, which eventually would be carved out of Tom Green. Indeed the *Live Stock Journal* of October 6, 1883, reported that Mayer "owns and controls forty-eight miles of range on the east side of the Pecos." The *Journal* further noted that he and Pulliam, who now controlled thirty-five miles of river on the west side, had a staggering total of 24,000 cattle on the Pecos. The two cattle kings were even considering a merger with a third party to establish a company with two million dollars in capital stock, a proposal which never materialized.

To work such a sea of cattle in rugged country required a large remuda, for each cowboy typically had seven animals in his "mount"—two drive horses, two roundup horses, two evening horses, and one night horse. In August 1883 Mayer dispatched Rufe Moore on a horse-buying trip through South Texas. In the first week of September, Mayer shipped two cars of animals out of San Antonio, likely the horses acquired by Moore and reserved for the JM or Circle Dot.

Up until now, Mayer had been content to graze half-wild longhorns on the JM and Circle Dot; they were hardy, resistant to Texas fever, strong enough to march a thousand miles, and able to produce calves at an advanced age. But he was also wise enough to appreciate the attributes of high-grade cattle, whose frames carried more meat, and of a more desirable quality, than the lanky longhorn. In short, crossbreds sired by blooded bulls—if they could endure a range as rugged as the JM or Circle Dot—would mean increased revenue.

High-grade bulls and the harsh West Texas landscape were about to collide head-on.

Chapter 6

Building a Quality Herd

In 1890, seven years after Mayer decided to introduce blooded stock to the Pecos, the *Cheyenne Live Stock Journal* expressed the attributes of high-grade bulls in practical terms: They increased the weight of steers in a herd by 150 to 200 pounds each at maturity. Coupled with the "finer form" of the cattle, the increased weight translated into an additional fifty cents to one dollar for every hundred pounds at market time. "This means," noted the periodical, "anywhere from $10 to $20 per head advantage by reason of a pure-bred sire."

A precedent already existed for blooded cattle on the Pecos. Sometime before the fall of 1880, Lewis Paxton and Milo L. Pierce introduced Durham bulls to their twenty miles of river frontage just above the New Mexico line. Still, high-grade animals were an unproven commodity on the Pecos, and it took courage for Mayer to invest money and effort in the project.

His first step was to board a train for Chicago in search of high-quality bulls, a task he probably would not have entrusted to anyone else. He selected thirty-three Hereford yearling bulls from D. C. Paxson and Company, which shipped the animals for Colorado City, Texas, on November 12. Mayer then traveled to the Kansas City Fat Stock Show, where he purchased more than 100 blooded calves, including 54 Hereford bulls, 17 shorthorn bulls, one Polled Angus bull, and four or five Polled Angus-shorthorn crossbred bulls. Additionally, he acquired as many as 83 shorthorn heifers and five Hereford heifers. The animals, for which he paid a "handsome sum," reported the *Texas Live Stock Journal*, were from herds in Illinois, Canada, and Kansas. Mayer also purchased another 21 Hereford bulls under circumstances that are unclear.

Mayer then took a train to Fort Worth and waited for the Fat Stock Show cattle to catch up with him. They arrived on November 17 and stirred up considerable interest in the *Live Stock Journal*, which was published in Fort Worth. "We . . . can truthfully say a nicer lot of plums would be hard to find," said the November 24 edition. "We were convinced, after seeing a hornless, five-months-old heifer by Polled Angus bull out of a Shorthorn cow, that no better specimen of a perfect calf for the Texas range could be found; indeed, the entire list was most pleasing to the eye. Form, substance, quality and thrift were all evidenced in the calves."

The Kansas Fat Stock Show cattle, as well as the additional Hereford bulls, were shipped west on the T&P Railroad to Colorado City. There, Mayer sold two Polled Angus bulls to Z. Smissen, then his drovers pointed the rest of his blooded acquisitions south toward Grierson's Spring, 125 miles distant.

Grierson's, situated about sixteen miles southwest of present Big Lake and twenty miles east of Pontoon Bridge, offered perennial water, a rarity in this thirsty country. The U.S. Army had established a military camp here in 1878, and the Upper Road from the Middle Concho had been rerouted to pass through the site en route to Pontoon. On June 1, 1882, the State of Texas deeded the land to J. P. Hodgson, but military officials continued to garrison the camp.

The army abandoned the grounds in late September of 1883, and perhaps Mayer had been privy to its rumored closing through his association with Camp Peña Colorado. Regardless, he obviously moved quickly to secure temporary control of the two sections of private land that comprised Grierson's, no doubt in anticipation of his plans to bring in blooded stock. The fact that open state land surrounded the oasis doubtless influenced his actions.

Mayer considered his imported cattle the best that had ever

entered Texas, reported the December 15 *Texas Live Stock Journal*. Dreaming of breeding the best class of cattle known, he planned to run his finest JM cows with his new bulls. "He has the native and acclimated stock to breed up on," observed the *Journal*, "and intends sparing no expense to make his business successful." Even as Mayer returned to San Antonio after a two-month absence, parties already were applying for the first stock of calves sired by the bulls.

A few Pecos cattlemen weren't so sure about Mayer's grand scheme. Three years later, after a couple of trials by fire on the river, Pecos cowman W. W. Peary blasted the idea of blooded cattle in the region. From his vantage point as foreman of the enormous TX outfit on both sides of Horsehead Crossing, Peary believed that native longhorns, which could fend for themselves, were the ideal cattle for West Texas.

"Grade [blooded] stock do not rustle worth a cent," he said, "and the introduction of the thoroughbreds into the range districts was the death knell . . . . The longhorn animal can take care of himself anywhere, and when you run them together on the range, you can readily detect their great superiority over the grades. The longhorns are a distinct class of cattle in themselves, with a physical structure peculiar to the surroundings."

Throughout the winter of 1883–84, Mayer held his prized cattle at Grierson's sweet waters and ample forage, letting them acclimate and mature a little before casting them to the whims of the Pecos. The following spring, his cowboys drove the animals down the relatively new road for Pontoon and turned them loose on the JM.

Despite occasional criticism by cowmen rooted in tradition, Mayer's progressive approach to his cattle business paid huge dividends. In succeeding years he continued to buy top-notch bulls, reaping the profits from their calves. "It is just as easy to produce this class of cattle as an eight-dollar yearling," the *San Antonio Stockman* paraphrased him in April 1897. "Buyers will hunt you up for the well-bred cattle, while the scrub is a drug on the market."

By the early twentieth century, not only was sixty percent of the JM herd purebred Hereford and the remainder Hereford-longhorn crosses, but the finest Hereford cattle in West Texas grazed Mayer's new Quien Sabe range. Moreover, the skeptics finally had come around to his way of thinking.

"'Better bulls, high grades, richer blood, shorter horns, deeper colors' is now, and has been for several years, the cry," said stockman A. S. Gage of Pecos. "And the reward is even now making itself manifest."

Mayer was also a careful range manager, at a time when ranchers routinely stocked the Pecos heavily without giving a thought to overgrazing. Not until August 1884 did stockmen issue an official warning against locating additional herds on the New Mexico stretch. "[It] is now greatly overstocked from the Texas line to its source, particularly [in] the present season, owing to the great drouth," said the resolution by Lincoln County Stock Association. "[There are] hundreds of dead cattle strewed along through the trails and ranges, from one end of the county to the other."

Mayer long before had recognized the range's limitations, prompting him to institute relatively prudent grass management. Already, he seemed to realize that the only thing predictable about Pecos droughts was that if there wasn't one this year, a worse one probably waited the next. Among the first cowmen to discern the dangers of overstocking, he always tried to secure enough grass to nurture each animal year-round, prompting the *Stockman* to declare it as "another reason for Mr. Halff's success in the cattle business."

Furthermore, as much as time and logistics allowed, Mayer was a hands-on owner. In the spring of 1884, F. J. Arno gained fascinating insight into the intensity of Mayer's involvement. With roundup approaching, Mayer found the JM in need of thirty-five to forty saddle horses younger than eight years of age. He offered Arno a commission of five dollars for every suitable animal he could locate in South Texas. Within a fortnight, Arno penned a small herd of cow ponies at Hondo City and contacted Mayer, who accepted thirty-three and shipped them to Marathon.

A few days later, Mayer secured Arno's services to accompany him to the JM. After a twenty-hour train ride from San Antonio, they unboarded at Marathon, where Circle Dot cowboys outfitted them with a buckboard and team for the short trip to Peña Colorado. Near the post, Mayer owned a small building in which a lessee operated a store and cantina that catered to soldiers and cowboys. That evening, as Mayer and Arno settled in for a night's sleep at the nearby Circle Dot headquarters, they were shaken by an uproar at the cantina. Related Arno:

> It seemed that hell broke loose, from the shooting and yelling that followed, but as they had lots of room, no one was hurt. And as they had no law, they kept on until they were tired. It made old man Halff a little nervous, but it quieted down about 11 p.m. and we got to sleep.

It would not be the last time a ruckus erupted at the Halff building; years later the establishment would prove troublesome to Camp Peña Colorado's commander and an embarrassment to Mayer.

For the time being, however, Mayer was more concerned with ensuring that his recent shipment of horses to Marathon reached the JM, 125 miles away by trail. The animals were already en route for Lancaster Crossing on the Pecos, and he wanted to overtake the drovers. Early on the morning after the cantina incident, he and Arno struck out for the JM in a buckboard pulled by two mules whose diminutive size belied their spirited natures. With Arno at the reins trying to hold the animals in check, the two men covered the seventy miles to Fort Stockton by sundown. There they caught up with the horses and stayed over for a day, allowing the mules to rest and Mayer to conduct business. Meanwhile, his drovers pushed ahead with the herd, carrying a request from Mayer to have a wagon and team waiting for him at Pecos Spring; the river was up and he would not be able to cross in his present rig.

The next morning, with the help of a half-dozen Mexicans, Mayer and Arno hitched the anxious mules to the buckboard and rumbled toward the Pecos. The road was rough but the mules more than willing, and before sundown Mayer and Arno reached Pecos Spring to find the expected wagon and team. While Arno unhitched the mules and prepared to stay the night, Mayer switched wagons and continued on to the JM. Two days later, his business completed, he returned to Pecos Spring and again changed vehicles. By now the mules were gentler, subdued by the hard journey from the Circle Dot, and Arno had no trouble at the reins as they headed out for Haymond, a railroad station sixteen miles southeast of Marathon. The men were forced to make their own trail through rugged canyons and over steep hills that demanded daring and initiative.

"Many times," recalled Arno, "I had to take a rope and tie our hind wheels to the body of the buckboard in place of a brake, and do the sliding act to the bottom of the hill."

The first night away from the Pecos they had to make dry camp; the mules went thirsty and the men had to ration what remained in their two-gallon canteen. The next morning, they finally struck water in a canyon, where they allowed the mules to drink and graze until after lunch. Forging on into evening, they pulled into Haymond, where Mayer arranged for his rig's return to Peña Colorado while he and Arno embarked for San Antonio by train. Mayer arrived home "a little worn," recalled Arno, "but still in the ring."

Decades later, Arno reflected on their whirlwind journey, one not unlike many that Mayer undertook on a regular basis. "That was one of the hardest driving trips in my experience," Arno recalled. "We made a circle of nearly two hundred miles over very rough country at a time [when] there were few or no roads and no settlements except in the towns."

A few months earlier, Mayer had contracted with Joe Carle of Castroville and Bill Foster, Medina County sheriff, for spring delivery of 1,500 Carle-Foster cattle. Mayer, in turn, was to supply the mixed herd to the firm of Ward and Courtney for a new ranching operation in Duncan, Arizona. Now, in the wake of spring roundup, the Carle-Foster cattle awaited pickup. Mayer, likely impressed with Arno's performance of duties west of the Pecos, employed Arno to class the cattle prior to acceptance.

Mayer continued to expand the JM range in the spring of 1884. On March 28 he purchased 640 acres on Live Oak Creek from T. P. and Annette Moore of Jackson County, Kansas, for $800. The 640-acre tract, situated seven and a half miles northeast of old Fort Lancaster, gave the JM an invaluable water source ten miles back from the Pecos. Furthermore, the location provided for easy wagon access, as it lay along the road from Fort Lancaster to Grierson's Spring.

The JM was still an open domain at this time, even though a legislator by the name of Patton commented early in 1884 that the Pecos now had eighty miles of wire fence. Patton's remark was so inaccurate that it moved Mayer, in the *Texas Live Stock Journal*, to offer the representative five dollars a day for life if he could produce the alleged fencing.

In late April, Mayer increased the numbers of cattle roaming free on the JM by bringing in 2,500 young steers and heifers from Del Rio, a town situated on the Mexico border to the southeast. As usual, he received and shipped the animals personally. Once the cattle were on the JM, he planned to cull the heifers for range use and put the steers into a northbound herd. The steers may have formed part of the 3,000-strong drove he dispatched for Wyoming that year in the charge of Ike Hill, for a convenient cattle trail to points north struck the Pecos just upstream of the JM.

Dubbed the Goodnight-Loving Trail by later chroniclers, the route originated in the Brazos River country near Young County and hooked south and west to terraced Castle Gap and beyond to Horsehead Crossing, the most infamous ford in the West. From the JM's north boundary, it would have been only a day's drive to the crossing. Once there, drovers could turn their herd onto the dusty trail and strike out upstream through the barrens. For a hundred miles or more, the trace held to the east bank (avoiding only major bends) before fording at either Narbo's Crossing, northwest of present Mentone, or Pope's Crossing, a mile below the New Mexico line. Above modern Carlsbad, New Mexico, herds again gained the east bank, which they hugged all the way up through Fort Sumner. Two or three days beyond, the trail finally left the Pecos and pressed on for Colorado and northern territories.

Despite the Montana-bound herd, Mayer still believed the future of transport lay in railroads. Even as his drove marched north in mid-June, cowboys at Colorado City, Texas, loaded thirty rail cars with Halff cattle for shipping to market on the T&P.

By early September, plans were underway to form a Pecos Valley stock association, for the river was alive with cattle. The JM itself had a reported 1,000 head in Tom Green County and a declared 2,500 head worth $30,000 in Crockett County. In the state as a whole, 5.5 million cattle grazed the range, even after 5.4 million had marched north the past nineteen years. The Texas Panhandle, almost fenceless, boasted an enormous number of beeves, as did the open ranges of western Indian Territory and southwestern Kansas. Throughout those north-lying regions, there was nothing to hold a herd in check except line riders or an occasional drift fence, including one that spanned the Panhandle north of the Canadian River. By tradition, ranches relied on general roundups to reclaim drifted cattle, a system which heretofore had been generally successful. Nevertheless, with a bovine prone to flee an oncoming winter storm, the entire region from southern Kansas to the Mexico border was primed for disaster as it entered the winter of 1884–85.

Late in 1884, a vicious blizzard roared out of the northeast and struck the South Plains, setting more than 200,000 cattle marching before a howling wind. The relentless storm persisted for a week, shrouding the plains with snow, and cattle from the Cimarron and Arkansas rivers piled up against the Canadian drift fence and died. Thousands of others bridged the fence by climbing the carcasses and pushed on madly, an incredible force impossible to hold back. Julius Henderson, at his father's ranch eight miles north of Odessa, watched a mile-wide herd thunder past in a run, trampling out a great trail in an eighteen-inch snow. The seething mass, only a part of the Big Drift, never paused until it struck the impassable Pecos. Many of the thirst-crazed animals plunged over the bank and perished, claimed by the river's treachery. The rest inundated the range from above Pecos City to the JM. "Cattle have drifted worse than at any time in the history of Texas," noted the *Texas Live Stock Journal* on March 28, 1885. The bad man's hell had become a cattle hell, and for cowmen like Mayer, things would only get worse.

The storm passed, but now the JM beeves and other natives were left to compete with the invaders for winter forage. One outfit alone, the Quien Sabe above Horsehead Crossing, reeled under the impact of 10,000 drifted animals. Soon the tufts of grass up and down the river were mere nubs, crushed under the hooves of starving cattle growing thinner by the day. "If [the drift cattle] are permitted to stay," lamented the *El Paso Times*, "they will soon eat out all the grass on the Pecos." Nevertheless, the fragile condition of the migratory beeves precluded any notions of an early return to the South Plains.

By early 1884 the grass nubs had turned to dust, laying bare a swath ten to fifteen miles wide all the way downriver to Pontoon Crossing. The migratory cattle in particular dropped by the thousands, casting a stench across the land. The carcasses became so concentrated that one cowboy, Pat Wilson, reported that he could have walked five miles by stepping from one to another. Meanwhile, the sky swarmed with buzzards eager to feast, and the river below was so infested with carrion and maggots that another cowhand, H. M. Hill, described it as "absolutely dammed"; he could have spelled it either way.

Still, there were tens of thousands of foreign cattle still crowding the range, and South Plains and Pecos outfits alike faced a daunting task. "It will require the combined efforts of the entire western country to get the cows home to their respective ranges," observed the *Texas Live Stock Weekly* on March 28.

In mid-spring, with the general roundup still weeks away, a few Pecos outfits managed to cull enough of their respective stock from the packed range to put together herds. In late April, the TX Ranch at Horsehead Crossing started 3,100 graded Durhams and longhorns on the trail for Indian Territory. Simultaneously, Mayer's cowboys were evidently working the JM range in preparation for rail shipments.

On May 18 a train load of Halff beeves rolled into National Stock Yards in St. Louis, the first shipment of Texas range cattle to reach the city that year. The fact that the animals were described as in only "fair flesh" and "coarse" suggests that they were from the beleaguered JM. The cattle, which averaged 863 pounds, sold for $4.00 to $4.15 per hundred pounds. A few days later, another train load of Halff beeves did better in Chicago, netting $4.55 per hundred-weight at Union Stock Yards.

Additionally, sometime in early spring Mayer dispatched Monroe Hardeman north with a herd of steers, possibly bound for Montana. Considering the early start and the poor condition of JM stock, it is likely that the drive animals carried the Circle Dot brand. The drove, tracing the Goodnight-Loving Trail, passed Fort Sumner May 4 and reached the Colorado line approximately three weeks later. Somewhere in Colorado, the outfit met with trouble; a mob of local cattlemen, perhaps protective of their range, attempted to turn the herd back. By now Mayer had joined the drove; he may have accompanied the outfit from its early stages. Whatever the nature of the mob's complaint, Mayer was successful in pushing his herd on through to its destination.

Perhaps Mayer had anticipated the Colorado problem, which could explain his absence from the JM at a time when the largest roundup in western history was taking place. It began about early February, when hundreds of cowhands from northerly regions converged on the Pecos and began cutting the Pecos stock from the much larger Plains herds. The operation involved at least twenty-seven wagon outfits, each with up to twenty cowboys, as well as 1,000 saddle horses. "The whole world," observed cowhand Lod Calohan, "was working cattle."

Within a month they had pointed 25,000 foreign animals for their home ranges. But the number of cattle yet to be worked was staggering, and time was an enemy in this land of inadequate forage. The cowhands' efforts were buoyed on May 20, when the Pecos outfits presumably rendezvoused as planned at Pontoon Bridge and set to work. Superintendent of the east-side roundup, which included the JM country, was Barnes Tullous of the Quien Sabe, who in another decade would figure prominently in Mayer's cattle business.

With grit and persistence, the combined forces pushed an astounding 120,000 to 150,000 cattle back toward their native ranges by July. Tens of thousands more had perished on the Pecos, their carcasses left to fester in the summer sun. When Mayer returned from Colorado in August, the Big Drift was finally just a memory, but in its wake loomed a blight even more sinister.

Chapter 7

Drought Woes

When a general rain muddied the bare alkali of the ravaged Pecos on June 5, 1885, JM cowhands were too concerned with ridding the country of drift cattle to take notice. Even out in the Big Bend, where showers were always the exception, the downpour probably evoked no more than passing gratitude from Circle Dot hands conditioned to a blazing sun.

Little did any of them know that it would be the last fruitful rain to grace the region for almost the next two years.

Mayer's decision in early fall to expand his grazing lands to Doña Ana County in New Mexico Territory was probably more than fortuitous timing. His instincts probably told him that the grazed-out JM would take several seasons to recover, and that by widening the scope of his operation, the impact of future calamities would be lessened. At any rate, on September 25 he purchased 120 acres on the south end of the Organ Mountains from John L. and Margaret Cook of Harrison County, Missouri, for $1,643. The Organs, twisted and spired, rose up 9,000 feet in majestic desolation in south-central New Mexico and stretched twenty miles from north to south, broken only at San Augustine Pass. Yucca, thornbush, and cacti were ever-present, but so were grasses at the range's base and in its chiseled canyons.

Mayer's acreage, situated east to southeast of Las Cruces,

53

comprised three plats: the first on the north end of Rattlesnake Ridge; the second on Boulder Canyon's west side; and the third near the juncture of Finley and Long canyons. Almost assuredly, each site included water, allowing for the possibility of controlling the bordering open range.

Mayer took on a partner in the Organs, a miner named Sam Taylor who was known for his enterprise in developing the range's water sources. Earlier in the year, Taylor had laid pipe from an Organ Mountains spring to a series of watering troughs for stock. Within a month of entering into partnership with Mayer, Taylor secured Lew Kelsey to dig a well.

Mayer was not content to limit his New Mexico interests to the Organs, however; by early December he was ranging thousands of cattle across his new lease in the Brazito Grant. By July 3, the *Rio Grande Republican* was hailing Mayer, along with Benjamin Davies, as a "cattle king" of the Organs' east side. The extent of Mayer's Doña Ana operation is reflected in the May 1886 tax roll, for which he declared personal property of $30,750, a total which suggests approximately 3,000 beeves. Many of them presumably bore a JM brand.

Rachel, Mayer's wife, was seriously ill in the fall of 1885, which limited Mayer's ranching activities. The *Texas Live Stock Journal* reported in late November that she was out of danger, but Mayer probably stayed in San Antonio several weeks more while she convalesced. By January Rachel apparently had recovered, and Mayer resumed his battle with drought.

When he attended a Texas cattleman's association meeting in Austin on January 12, 1886, stock on the Pecos were still in good condition. Indeed, the drought did not even merit a mention in a *Texas Live Stock Journal* account of the meeting's proceedings. As months passed and the sky remained barren, however, a withering pestilence began to creep across the land. The JM range and those of other east-side Pecos outfits were especially hard hit, for the drought's quick onslaught had given the grasses no chance to recover from the Big Drift. The Big Bend, always shy of water, was almost as vulnerable. Before the year was over, recalled one pioneer, a man could ride the Rio Grande from Boquillas to Santa Elena Canyon and never find a puddle from which to drink.

Mayer, however, forged onward in the evident hope that spring rains would replenish the region. He maintained his Circle Dot herd of 20,000 head, which now consisted primarily of graded Durhams. In early February Mayer was a delegate to the International Range Cattle and Horse Growers' Association convention in Denver. The next month, he began buying cattle for a drive, perhaps making at least one concession to the possibility of inadequate forage on the trail—he wanted no steer under two years of age.

Sometime that spring, two Halff herds embarked from Peña Colorado, at least one of them bound for Montana with drover William Doak. Problems brewed from the start, for New Mexico stockmen had issued a warning against driving through their ranges in this droughty season. "Cattlemen of that Territory say [it] must not be done, or trouble will surely follow," reported the *Globe Live Stock Journal* on April 27. "They say their ranges are fully stocked and that there is neither grass nor water for transient stock." A subsequent rumor had it that an armed force waited in Trinchera Pass in northeastern New Mexico to deny the passage of herds.

Although the threatened confrontation never materialized, nature's wrath was less obliging. The Big Dry followed Mayer's droves up the Pecos into New Mexico, torturing the animals and exacting an enormous toll. Some trail outfits, in fact, had to turn their herds loose on foreign range, for the malnourished cattle could trudge no farther through the dusty hell of sun-baked alkali that defined the Pecos in '86. All the way up through Fort Sumner, a miles-wide swath of barren ground crept along at river's flank. Soon the *Weekly Yellowstone Journal and Live Stock Reporter* of Miles City, Montana, would lament that "the loss of cattle on the trail the present season . . . is without parallel in the history of the range cattle business. The range is strewn with carcasses. . . . The occupants of the range along the trail have thrown every obstacle in the way, and the owners of herds have found difficulty in crossing every political boundary line."

Mayer's cattle kept up their march, however, the carcasses of the weaker animals marking the trail behind. Evidently in deference to the reported militia in Trinchera Pass, Mayer or his trail boss devised an alternate route to reach Colorado with his larger herd, which had tallied 2,500 head in Texas. After leaving the Pecos above Sumner, the drovers pointed the animals eastward into the Texas Panhandle to avoid Trinchera Pass. At Tascosa, in present Oldham County, they struck the Dodge City trail and traced it north until veering for Colorado. Dying animals dropped with every passing mile, their skin drawn tightly over their frames in gruesome reminder of nature's sovereignty. By the time the drove passed Trail City, Colorado, sometime before July 11, the trace back to the Circle Dot was lined with 1,100 carcasses.

A Trail City correspondent for the *Texas Live Stock Journal*, reporting the herd's arrival, put Mayer's losses in perspective: He had lost forty-four percent to starvation. Meanwhile, said the correspondent, "those that pulled through are thin, but improving rapidly."

The drovers with Mayer's second herd, including Doak, avoided the same degree of disaster by combining trail with rail. They held to the established Goodnight-Loving route up past Fort Sumner, doubtless suffering losses, before setting out across the New Mexico plains for a railroad at Springer. Shipping the cattle via rail for Colorado, they reached Denver with about 2,000 head. From there the animals took to the trail again, all the way to Mayer's Wyoming range seventy-five miles north of Cheyenne.

Back in Texas, conditions had become even more perilous for livestock. April in particular was severe. "During the past month," C. A. Hamilton observed in the May 1 *Texas Live Stock Journal*, "more cattle have died than during the entire winter. They literally starved to death, for want of grass." Late in April, Pecos Valley Cattle Association, as it was now known, even canceled its May general roundup, for the impoverished cattle were too weak to be worked.

John Simpson, inspecting his giant Hash Knife outfit on the Pecos in early May, was appalled by a country so mournfully dry. "I tell you, it is awful," said Simpson, whose west-side range, unlike the JM, had been spared the added devastation of the Big Drift. "Of course, there is plenty of water in the river, continuously running snow water from the mountains, but the country bordering on it for eight and ten miles back, is as barren of food, almost, as a sand bank. Many of our steers go back into the interior fifteen and twenty miles to where a sufficiency of cured grass can be had, but the amount of travel necessary to cover the distance between the elements of existence [i.e., forage and water], keep them poor and fit subjects of the bog [in the Pecos quicksands]."

A May 18 report by a Big Spring, Texas, resident in the *Globe Live Stock Journal* was even more bleak. "The plains west of here are parched and dry, and the carcasses of thousands of cattle are to be seen in every direction," he wrote. "A rough estimate places the rate of mortality by thirst and starvation at 900 head per day. Fully 20,000 carcasses cover the plains. The stench as one passes along the Texas Pacific west of here is terrible."

The onslaught of hot weather only made matters worse, as an El Paso correspondent indicated in a May 25 letter to the *Texas Live Stock Journal*: "From the Pecos River for 100 miles below and 200 miles above the Texas and Pacific Railroad [a stretch which included the JM], cattle are reported as dying daily by the hundreds, and if all reports be true, I may say by the thousands. . . . The cattle are too poor and weak to bear rounding-up or handling, as would be necessary to move them; besides, there is no grass or water to move them on. There seems to be no alternative but to leave the unfortunate animals to rustle as best they can where they are, and rely entirely for their existence on an early rain."

The next year, M. L. Liles gained insight into the critical conditions on the JM. Striking the Pecos at Pontoon Bridge, he found the range still devastated. "They hadn't had enough rain to wet a man's shirt sleeve for eighteen months," he recalled in 1937.

Upstream at Horsehead Crossing, the TX outfit did enjoy a "splendid" rain in early June of 1886, as did Presnall & Mussey's range in Pecos County. By early July, however, the beneficial effects of the showers had disappeared, and cattle along the Pecos resorted to foraging on mesquite beans, of which there was an unprecedented crop. Over the next two weeks, isolated rains west of the Pecos briefly lifted the spirits of cattlemen but accomplished little else.

In early summer Mayer trailed eight hundred cattle into

the Organs, probably from his beleaguered Texas range, and turned them loose immediately east of San Augustine Pass. With pasturage and water so coveted in the Organs, adjacent rancher B. E. Davies—likely Mayer's fellow "cattle king" Benjamin Davies —sought an injunction to keep Mayer from grazing and watering cattle on the Davies property. In proceedings before territorial district judge William F. Henderson, Mayer asserted that not only did he possess land and sufficient water in the area, but that it was customary in Doña Ana County to let cattle range at will. Nevertheless, reported the October 9 *Rio Grande Republican*, Judge Henderson sustained the injunction and ordered M. Halff & Brother to keep its cattle off Davies' premises.

A general roundup was finally held in the Big Bend about September, but it only served to wreak tragedy much greater than any cattle loss. The roundup crew, bossed by Billy Kincaid, included cowhand Jim Davenport and Circle Dot foreman Elkanah M. Herreford, one of Mayer's most responsible men. A Missouri native, Herreford was seventy-one and saddled with the nickname "Old Man Herreford," but his age did not deter him from taking an active role in the roundup.

According to Joe White of the Circle Dot, the incident began when Kincaid assigned Davenport to hold a herd along a stream. As the cattle watered and grazed, Davenport fell asleep under a shade tree, allowing many beeves to wander away. Herreford apparently was in the area, and he may or may not have reported the cowhand's incompetence to Kincaid. Either way, when Kincaid fired Davenport over the incident, Davenport blamed Herreford for his ill fortune.

White's account holds that Davenport rode straight to a Marathon saloon and bellied up to the bar; when Herreford walked in some time later (one version says days later), Davenport confronted him and shot him dead. The October 2, 1886 *San Angelo Standard* placed the August 6 killing at the Circle Dot, which the *Standard* termed the "Herreford ranch."

According to the newspaper, Texas Rangers soon tried to arrest Davenport but the cowhand resisted, wounding two rangers and sustaining a severe wound of his own. The shootout reportedly occurred at G4 headquarters at the northeast base of Packsaddle Mountain, thirteen miles north of Study Butte.

Stifling the bloody flow in his side with a handkerchief, Davenport fled on horseback for Mexico, accomplishing three hundred miles before stopping. By early October, however, the "bold, bad man," as the *Standard* described him, had been captured and confined to jail in El Paso.

Convicted, Davenport served five years for the murder, according to White. Upon his release, he reportedly swore vengeance on Mayer, whom he believed was instrumental in his conviction. Mayer probably breathed a little easier one day about 1899 when lawman Will Wright killed Davenport in the South Texas town of Cotulla.

Herreford was buried just south of the mountain pass near Peña Colorado, his tombstone a tragic monument to drought's ability to destroy lives, both animal and human. By late October, Mayer had replaced him as Circle Dot manager with S. P. Pulliam, brother of Pecos River rancher Nub Pulliam. The new manager's tenure was evidently brief, for Mayer transferred his JM foreman, Jim Cline, into the position later in 1886. Cline would remain Circle Dot manager for approximately a year.

As the drought wore on into another winter, Mayer sought additional grazing for his JM stock, which were not holding up as well as Circle Dot cattle. Just north of Pontoon Bridge, the State of Texas controlled a broad band of university land along Five-Mile Draw, and Mayer evidently had his eye on it. On December 31, the *San Antonio Light* announced that he had filed application with the land board for the lease of twenty university sections. Although the state evidently denied his request (not unexpectedly considering the unspecified conditions Mayer attached, said the *Light*), this early interaction with the board paved the way for future negotiations.

Mayer was more successful in arranging a trade for the H. H. Carmichael and Company Ranch in Bandera County, cradled in the hills northwest of San Antonio. The *Dallas Morning News* announced the $27,000 purchase on January 5, 1887. Carmichael, the leading stockman and merchant in Bandera County, would later join Mayer in a Big Bend ranching venture.

During the second week of January, Mayer attended the fifth annual State Live Stock Association convention. By then the drought had wreaked havoc with the cattle market, and the executive committee acknowledged the "hard times and pressure" faced by stockmen. "Two years ago, average stock cattle were readily sold in West Texas at $20 per head," the committee reported. "Today the same class of stock cannot be sold for more than $10." Despite the grim situation, the committee took heart, noting that Texas ranchers had showed their "staying powers" to the world at large.

Endurance and mettle were qualities that cowmen like Mayer needed in 1887, for the drought showed no signs of breaking even as spring approached. Mayer's reaction was to extend his grazing lands even farther westward, almost to the Arizona line, and north into Indian Territory.

A couple of weeks after the February 2 wedding of his nineteen-year-old daughter Hennie to New York native Frederick Goldsmith in San Antonio, Mayer took a train out to Peña Colorado. Loading 1,000 stock cattle into rail cars at Marathon, probably in March, he shipped them west to Separ, New Mexico, a T&P station southeast of Lordsburg. It was not his first business concern in Separ; a few months before, he had dispatched Jim Cline to the site to deliver cattle to a buyer. This time, however, Mayer arrived as a partner in a nearby ranch established by J. B. Weems. In exchange for a half-interest in the land, Mayer gave Weems a like interest in the herd.

The Weems-Halff operation was located north of Separ in the south end of the Burro Mountains. The Burros actually consisted of two parallel ranges, the Big Burro and the Little Burro, separated by the Mangas Valley. The smaller range, predominantly bare and only eight miles long, rose to 6,396 feet a few miles southwest of Silver City. Its forested sister, now in Gila National Forest, jutted to 8,045 feet and stretched more than twenty miles from north to south.

Even as Mayer made inroads in the Burros, his cowhands delivered 1,000 steers by rail to A. Mills in the Cherokee country of present Oklahoma. Simultaneously, cars bearing an additional 1,100 steers rolled into Indian Territory on their way to yet another of Mayer's new ranges. This range apparently lay in the Cherokee Strip, a 58-mile-wide band of country (centered on the Arkansas and Cimarron rivers) that stretched between the ninety-sixth and one hundredth meridians. In 1883 the

Cherokee Strip Live Stock Association, a cattleman's corporation, had secured a long-term grazing lease on the Strip from the Cherokees, and now the company sub-leased it to individual stockholders.

Mayer, after two months on his western ranches, finally returned to San Antonio in mid-April of 1887. The circumstances were tragic. Rachel had given birth the previous September 23 to a child, Sidney, and the infant boy had died on April 15. Mayer, despite his frequent absences, was by all accounts close to his family, and one can only speculate on his emotions at the loss of his son.

But out of tragedy rose hope, for almost simultaneous with Sidney's death came good news, the kind that could encourage a beaten cowman to throw himself again on the mercy of nature's inscrutable ways. On April 12 copious rains—the first rainfall of any kind in seven months—fell throughout the Pecos country and westward beyond the Big Bend. "Hope is again revived," proclaimed the *Texas Live Stock Journal* on April 16.

By mid-June, West Texas was truly alive once more. "The creeks are again running," reported the *Journal* on June 18, "the grass doing better than could have been dreamed of, good winter pasturage is assured, and the calf crop the largest . . . ever."

For some, the victory had come at a great price: Fully fifty percent of the cattle that had ranged along the Pecos two years before were now carcasses rotting in the late spring sun. "Texas cattle have passed through the darkest hour," observed the *San Angelo Enterprise*.

Nevertheless, added the periodical, for ranchers such as Mayer, "the bright sun of better days" was now emerging above the dark horizon.

Chapter 8

Home Life and Pecos Rustlers

The character strengths that guided Mayer in the cattle business carried over to his home life, in which he displayed love, caring, kindness, and occasional humor. Although he was away for extended periods, he seemed truly committed to Rachel and their children, leading the *San Antonio Express* in 1905 to term him "an affectionate man in his family circle."

For decades the Halff home was a center of social and cultural affairs in San Antonio, for Mayer possessed the unique ability to shed the trappings appropriate to a cow camp and assume the role of a highly refined gentleman. Surprisingly, his cowboy and cosmopolitan sides never seemed at odds. He neither inflicted a cowhand's undignified demeanor on San Antonio social circles, nor sought to impose on his uncomplicated cowboys the genteel ways of the city. Still, the well-bred San Antonians who graced his home would likely have been shocked by the crudities of a roundup camp, and his cowboys would probably have been even more stunned by the elegance of a San Antonio social event attended by people so prim and proper.

An inkling of the polish attached to an affair in Mayer's home is evident in a *San Antonio Express* account of a dance he and Rachel hosted for their daughter Lillie when she was nineteen. "This was a chrysanthemum party and the beautiful home was trimmed with autumn's green," reported the *Express*. "Even the incandescent lights peeped out from among white petals, and waxy amylax [sic] hung in festoons from the chandeliers. The table in the dining room was indented with fancy cakes and all manner of good things, and the silver and glass gleamed among the flowers and the greens."

Nor did any less glamour surround Lillie and the guests.

"The dark rich beauty of the young debutante," said the *Express*, "never showed to better advantage than on this happy occasion when she appeared in a gown of white crepe artistically combined with yellow. Pearls were her jewels and her flowers were yellow and white chrysanthemums. All of the guests . . . were decked out in the flower of the evening, and the ices were eaten from the heart of white chrysanthemums."

Even the entertainment revolved around chrysanthemums, with guests estimating the number of petals in a particular specimen. The prizewinner was Jesse D. Oppenheimer, who proceeded to win Lillie's heart as well; the two married on March 24, 1898, thereby joining two of the state's most prominent pioneer Jewish families. Jesse's father, Dan Oppenheimer, had immigrated from Bavaria in 1854 and had established a San Antonio dry goods business with his brother Anton after the Civil War. Operating under the name D. & A. Oppenheimer, they eventually had founded a bank in San Antonio and accumulated large ranching interests in West and Southwest Texas.

Lillie, apparently sheltered from the rugged life of her father's ranches, later imparted a sense of Mayer's refined side to her son, Jesse Halff Oppenheimer.

"He was very, very much of a continental gentleman, a person of culture and dignity," said Oppenheimer, who was born thirteen years after Mayer's death. "He was not rough-shod, like so many of the cattlemen were. He believed very strongly in education, and he ran a household that was very sophisticated and upgrade. He had the best of wines and the best of linens. He was the type of individual that could come back to San Antonio and be a cultured European gentleman, and then the next thing you'd know he was out at the Pecos or Marathon area again, being a boss of a cow operation."

Mayer was also a patron of the arts.

"My mother pictured him as liking music, and he encouraged my mother in singing and piano," related Oppenheimer. "According to her, he was a very adoring-type father and was terribly flattering to her. He doted on my mother; she was very much a favorite of his because she was attractive and musical."

Mayer also introduced his son Henry to the refined life, even as he groomed him for the role of a cowman who, like himself, knew the business from the ground up. An incident that Henry later related to his own son Albert provides a rare glimpse into Mayer's sense of humor in a cultured setting.

As Mayer had done for Lillie, he arranged piano lessons for Henry. Perhaps unaware that Henry did not share Lillie's musical aptitude, Mayer decided to showcase the lad's talent one day for visitors in their home.

"Son," he said, "play one of your pieces for the company."

"But I only know one piece," retorted the teenager.

With dry wit, Mayer made the most of an embarrassing moment. "Well," he jokingly chastised, "you didn't have to tell them."

Around San Antonio, Mayer was active in numerous civic groups and charitable causes. His social activities included membership in the San Antonio Club, which included an opera house. The April 1, 1887, *San Antonio Express* described the club as "an elevating and humanizing institution. . . . Within its ranks are to be found the men who have made San Antonio what it is to-day."

A man almost as busy with social affairs as he was with the cow business, Mayer nevertheless made time for extended vacations with his family in the 1880s. Leaving their children behind in 1881, he and Rachel sailed to Europe and visited his relatives, who had yet to meet her. Mayer's apparently poor health upon his return to America, however, led his sister Rosalie to pen him an October 19, 1881, letter urging him to relocate to Paris before he was "completely ruined . . . . Everything depends on your health."

Evidently regaining his prodigious energy, Mayer went on an ambitious trip to New York with Rachel the next year. In the summer of 1884, the couple carried their children on a two-month excursion to Detroit and the Great Lakes with Solomon and his family. Four years later, Mayer embarked on yet another lengthy tour of northern regions, likely in the company of his family.

Occasionally, circumstances prevented him from going as planned. Such was evidently the case on July 31, 1887, when Rachel, their younger children (Hennie was now married), and Solomon and his family left for New York without him. Likely, Mayer was consumed at the time with his new ventures in New Mexico, Indian Territory, and McMullen County.

Throughout his children's formative years, Mayer tried to give them all the advantages. He enrolled Alex in the respected German-English School in San Antonio, and in 1885 he introduced the sixteen-year-old boy to the dry goods business. Instead of starting his son in a managerial position, Mayer chose to have him work his way up through every department. Preparing Alex to succeed him at M. Halff & Brother, Mayer sent him to school in Cincinnati and to colleges in Boston and New York, a city where Lillie also would attend school. On January 23, 1899, Alex married twenty-one-year-old Alma Oppenheimer, the sister of Lillie's husband, further solidifying family ties between the Halffs and Oppenheimers.

Mayer also sent Alex's younger brother away to the best schools in the country. Henry attended Staunton Military Academy in Virginia and Eastman National Business College of Poughkeepsie, New York, from which he graduated August 27, 1893.

Rachel, meanwhile, stayed busy in societal circles and with charitable causes, such as her sponsorship of an 1892 ball for the benefit of the Red Cross. Like her husband, who once described San Antonio as "*the* town" for him, she seemed perfectly satisfied with city life, despite having the means to relocate to the country. Sometimes, however, she did venture away from San Antonio's cosmopolitan atmosphere.

One such occasion came immediately upon her return from a two-month trip to New York in early fall of 1887. Roundup season was drawing to a close, and Mayer was away at one of their ranches. With cattle business precluding his departure at the time, Rachel traveled out to join him.

In planning the roundup for the Big Bend that August, Presidio Stock Association had met in Murphyville (present

Alpine) and discussed the problem of illegal branding. In 1880s West Texas, it was a serious concern. With the range still open, a calf might well escape a branding fire. Unclaimed, the maverick, as it was called, became fair game for the first outfit to burn a brand in its hide. The indiscriminate branding of such a calf, done without consideration for who the owner might be, was known as mavericking.

Since unmarked calves often drifted onto neighboring ranges, mavericking often was abused. Some outfits had a "gentleman's agreement" not to brand mavericks except at general roundups—but gentlemen sometimes were few and far between on the frontier. In fact, for a dollar-a-day cowboy, it was only a small step from mavericking for his boss to mavericking for himself.

Such was the case with Henry Green, a cowhand on the JM and perhaps the Circle Dot. For years Green was a dependable "outside man," representing Mayer at area roundups and dutifully returning the Halff cattle to their home. His job often carried him to remote canyons and thickets where he saw numerous unmarked cattle ranging forgotten. Finally the temptation proved too great.

Green drew his wages at the Halff ranch, rounded up a few unscrupulous partners, and headed for the canyons and thickets with a branding iron. When mavericks grew sparse he took to brand-burning—altering an existing brand with a running iron. The new mark, never legally recorded, was known as a "slow brand." Mayer's TL brand, in fact, was burned into a TD by a party who never tried to claim it, remembered Big Bend pioneer Jim B. Wilson.

The Green gang's method of operation was to round up a hundred or so cattle at dusk and drive them under cover of darkness to a remote location. There they would hold them until the following night and resume the drive. When the herd was safely out of the area, the rustlers set to work with running irons, creating a brand unique to the region. The gang then freed the beeves, knowing that as soon as the brands healed, the rightful owners would consider them someone else's property. When the cattle went unclaimed at succeeding roundups, they would be considered strays, ready to be picked up and sold by the rustlers.

It became common knowledge that the Green gang was responsible for all the new brands on the range, but no one had enough evidence for a conviction. As J. Frank Dobie told the story, the Texas Rangers brought in an expert brand reader who determined that thousands of cattle were involved. Feeling the noose tightening, the rustlers fled the country. Green wound up in New Mexico, where in a bit of poetic justice he soon died in a gunfight.

It's not known how many unidentified brands were seen in the Big Bend roundup in fall 1887, but Mayer found his Circle Dot cattle in fine condition. He shipped 203 steers and 47 cows to Chicago from Toyah on about November 1, but the sale price was disappointing. Only seven weeks before, the *Texas Live Stock Journal* had bragged that cattlemen would "rake in . . . $4 per 100 pounds," but Mayer's steers, averaging 902 pounds, netted him barely more than half that amount. His Pecos cattle, meanwhile, benefitted from terrific early-November rains that filled outlying water holes to their highest levels in three years. Nevertheless, with cattle prices still down at the time of his late-December trip to Peña Colorado, he decided to hold his stock rather than ship.

In January a winter storm struck the South Plains, and though the consequences to the Pecos country were not as disastrous as the blizzard and big drift in 1884–85, they were nonetheless significant. Approximately 18,000 cattle swept down from the north-northeast, crossing the T&P track west of Midland before massing along the river. Exacerbating the problem was a series of blizzards in the JM region that month. Cattle losses were minimal, but the storms "scattered, drifted and mixed up the herds to a fearful extent," said the *New Mexico Stock Grower*. "It will take a world of labor and trouble to again classify them according to brand."

Evidently due to the continued low market value of cattle, many West Texas ranchers chose to ship beeves to Indian Territory early in 1888 and fatten them until the fall market opened. The railroad accommodated stockmen by billing shipments at a through-rate to Chicago, with a layover privilege of three months in the Territory. During one month alone, 30,000 West Texas cattle rolled into present Oklahoma, prompting the

Cuero (Texas) *Star* to declare that "the shipment of cattle from West and Southern Texas to the Indian Nation . . . surpasses anything ever known before in the history of the business."

Mayer contributed his share to Indian Territory, freighting more than 1,000 JM steers to the Cherokee country (for pasturing by A. Mills) in late February. Mayer followed up a month later with additional shipments, perhaps to his own holdings in the Strip.

In early April Mayer actually benefitted from the depressed cattle market, gaining a lower tax value for his Organ Mountains herds. Doña Ana County stockmen decided to assess their cattle at $8 per head, down from $12 the year before.

With rail cars booked for the foreseeable future by the unprecedented rush to Indian Territory, Trans-Pecos outfits were forced to send four or five herds, totalling 10,000 head, north by trail that spring. Although the drive was fading as a means of transport, stockmen unable to reserve cars could take heart. Of the 96,000 cattle that had reached Trail City, Colorado, in 1887, only 2,300 steers and 900 cows and heifers remained unsold.

In late winter of 1888, Mayer supposedly found a need for an additional dwelling on the Circle Dot and took steps to reclaim a building on the military reservation. Situated on the west side of the creek in Peña Colorado gap, it stood against a small mountain whose bluff formed one of its walls. For some time the structure had housed the store of J. J. Hess, who kept a small stock of groceries and beer. Mayer's real reason for evicting Hess may have revolved around that night in 1884 when gunfire at the establishment had kept him awake. Folklore even holds that soldiers killed an outlaw named Alexander at the site the same year.

On March 20, 1888, Hess requested permission from post commander E. G. Fechet to erect a new store building at a site selected by the army. The Adjutant General's Office, Department of Texas, had no objections, but left the decision up to M. Halff & Brother. The details are fuzzy, but Hess eventually vacated the premises without erecting a second building. Whatever the full story behind Hess' departure, Mayer did not follow through with plans for establishing a dwelling. Under different management, the store continued operation, setting the stage for an embarrassing series of events in 1889 and 1890.

By early April of 1888, Mayer had named Rufe Moore foreman of the Circle Dot and perhaps general manager of all his ranches. Mayer, now fifty-three, continued to stay personally involved, however, securing brief grazing rights in Pecos County for 1,100 cattle worth $7,700 and fifteen horses valued at $375.

He also extended his cattle operations into Bee County, just off the Gulf Coast in South Texas, where he had acquired four tracts totalling 2,856 acres from J. T. Byers. For the next thirteen years, Mayer retained this acreage while acquiring additional land in the county. Small by Mayer's standards—only 4,880 deeded acres at its peak—the Bee County outfit ran 150 to 350 cattle annually, along with a few horses.

In early May of 1888, Mayer traveled to Peña Colorado, where he found his stock in remarkable condition, despite a delay in seasonal rains. Soon he tested the Chicago market with twenty steers. The animals, which averaged 752 pounds, brought $2.60 per hundred pounds, up 45 cents from the previous November but still below a favorable price.

But better days seemed to be ahead. Plentiful summer rains spawned excellent grazing on the JM and Circle Dot, and when Mayer departed on vacation in mid-July, he left behind fat cattle "chewing their cuds on the green hills beyond the Pecos," said the *Texas Live Stock Journal.*

By the fall of 1888, West Texas had been free of drought for eighteen months. Except for stretches of the Pecos ravaged by the Big Drift, the range had bounced back remarkably. But an arid curse again descended from a cloudless sky and began to choke this thirsty land. It would not be as severe as the Big Dry of '85 to '87, but cattlemen would nevertheless feel its sting.

The dry spell went unnoticed initially, even though Big Bend cattle losses were up 70 percent and horses 25 percent in the winter of 1888–89. The *Texas Live Stock Journal* placed the blame solely on a reduction in grazing land due to a rapid "settling up" of the country. The effect of nesters on Mayer's Circle Dot business is not known, but by now a small settlement known as Buena Vista had sprung up a couple of miles south of Circle Dot headquarters.

Indian Territory continued to be the destination of choice for Texas cattlemen in the spring of 1889, with 143,500 cattle expected to reach the region by rail and another 84,000 by drive. In mid-February Mayer was in Fort Worth, perhaps reserving rail cars. He needed plenty, for he and Solomon planned to ship 5,000 cattle to the Territory via the T&P and GC&SF lines. The total may have included 2,100 steers, pastured primarily in Frio County, which the brothers delivered in March to Arkansas City Cattle Company in Panca, Indian Territory.

By May the Circle Dot and JM ranges were wilting under the impact of drought, for spring had brought only a lingering dry and torturous sun. The situation forced Mayer to lease pasturage near Sabinal in Uvalde County from a Mrs. Wish. He shipped two train loads of Circle Dot cows and calves to the acreage in mid-month and another train load by June 1. Even so, the Wish range couldn't absorb his entire breeding herd, leading the *Texas Live Stock Journal* to report on May 25 that he was considering stocking the "Kelly pasture" with an additional 2,000 Circle Dot cows and calves.

But even as Mayer struggled to save his breeding animals, he had to negotiate with Camp Peña Colorado on two troublesome matters far removed from San Antonio high society. They involved hogs, whores, and a hot-tempered saloon keeper—and Mayer was in for embarrassment.

Solomon Halff
— Courtesy Alex H. Halff

Rosalie Halff,
sister of Mayer and Solomon.
— Courtesy Alex H. Halff

Felix Halff
— Courtesy Alex H. Halff

June 16, 1856, advertisement in
Liberty Gazette
for Mayer Halff's store.
— Courtesy Alex H. Halff

Miscellaneous.

NEW STORE! NEW STORE!!
A. & M. HALFF have JUST OPENED, in a part of Judge Branch's Store, a large stock of SPRING AND SUMMER GOODS of the latest styles. Among their stock will be found Silks, Satins, Muslins, colored, white and embroidered, Bereges, plain and colored, Prints, Calicoes, &c. &c. Mens' and Boys' Ready Made Clothing, Boots, Shoes, &c. Also a large stock of worked Cambric and Swiss Collars, Chemizettes and Sleeves; Ladies' and Misses' Bootees, Shoes and Slippers, &c.

Our motto being "Cheap for Cash," we will not be undersold by any store in the place. ☞ Call and examine our stock. No trouble to show goods.

☞ The highest market price paid for HIDES, DEER SKINS AND BEESWAX. ap28

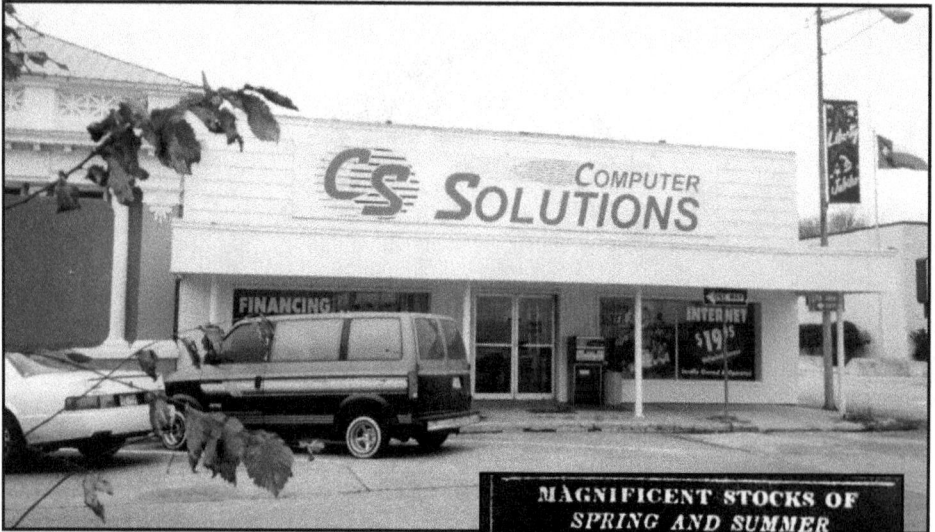

Mayer Halff's former store building in Liberty in late 1990s.
— Courtesy Alex H. Halff

Liberty Gazette ad of 1859 for Mayer Halff's store.
— Courtesy Alex H. Halff

M. Halff &
Brother store
(right) on
Commerce Street
in San Antonio.
— Courtesy Alex
H. Halff

Interior of M. Halff & Brother store in 1914.
— Courtesy Alex H. Halff

Rachel Halff.
— Courtesy Alex H. Halff

Rachel Halff at age twenty-six.
— Courtesy Alex H. Halff

Mayer Halff's one-time home on Goliad Street in San Antonio.
— Courtesy Jesse Halff Oppenheimer

Mayer Halff's initials at the Goliad Street home.
— Courtesy Alex H. Halff

The old Circle Dot and Gage country stretching toward the cone of Santiago Peak.

— Photo by author

Rainbow Cliff at Peña Colorado.

— Photo by author

Camp Peña Colorado in the early twentieth century.
— Courtesy Travis Roberts Jr.

James Travis Roberts (center) at former Camp Peña Colorado circa 1911.
— Courtesy Travis Roberts Jr.

Cowhands at Peña Colorado Creek.
— Courtesy Travis Roberts Jr.

The creek inside
Peña Colorado gap.
— Photo by author

Circle Dot branding iron.
— Photo by author

*Site of the Halff
store/saloon in fore-
ground beside road at
Peña Colorado.*
— Photo by author

Close-up of the Halff store/saloon site. The bluff at left formed one wall.
— Photo by author

*Ike Roberts and Travis Roberts Jr.,
grandsons of Circle Dot cowhand
J. J. Roberts, inspect adobe melt-down
at site of Circle Dot headquarters.*
— Photo by author

Maravillas Creek at Rock House Gap in Circle Dot Country.
— Photo by author

Circle Dot line camp in Rock House Gap as seen in 1999.
— Photo by author

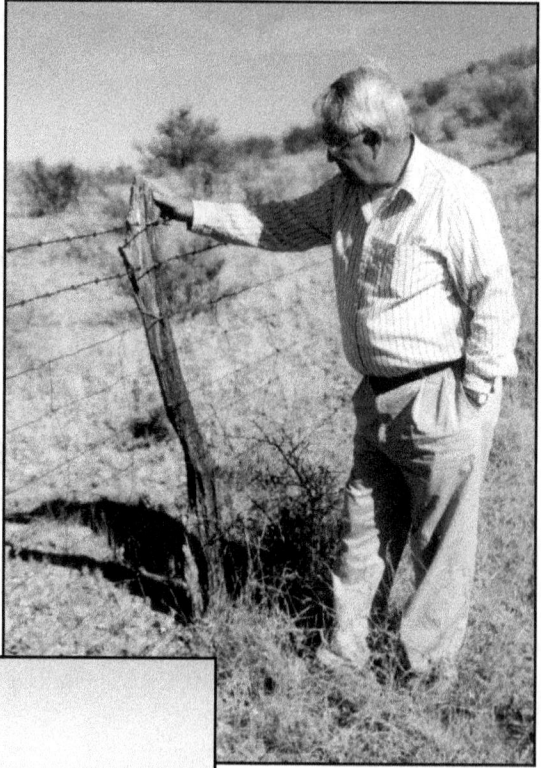

Travis Roberts Jr. inspects original 1886 drift fence in Circle Dot Country in 1999.
— Photo by author

The 1886 drift fence as seen in 1999. The Circle Dot was on the right, the Gage on the left.
— Photo by author

Original post, staples, and probable wire of 1886 drift fence in Circle Dot country as seen in 1999.

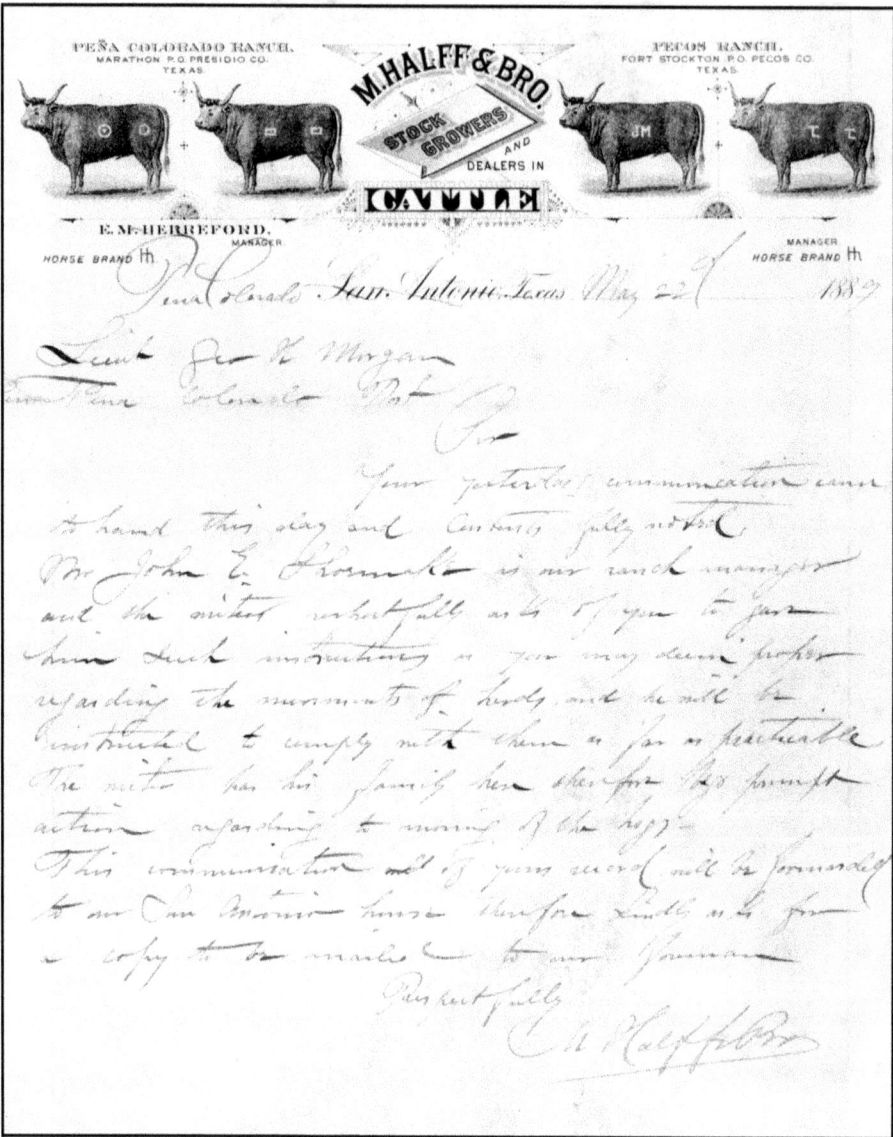

*Mayer Halff's May 22, 1889, letter to Lieutenant George H. Morgan,
showing M. Halff & Brother's letterhead and brands.*

— National Archives

Cowhands at an early Big Bend roundup.
— Courtesy Travis Roberts Jr.

*Cowboys around a chuck wagon in
the Big Bend.*
— Courtesy Travis Roberts Jr.

*Elkanah M. Herreford's grave near old
Circle Dot headquarterss.*
— Photo by author

STATE HISTORICAL SURVEY COMMITTEE

TEXAS

MARATHON

FORT PEÑA COLORADO, THE LAST ACTIVE FORT IN THIS AREA, ON THE OLD COMANCHE TRAIL, ABOUT 4 MILES TO THE SOUTHWEST WAS ESTABLISHED IN 1879.

MARATHON WAS FOUNDED IN 1881. NAMED BY AN OLD SEA CAPTAIN, A. E. SHEPARD, FOR THE PLAIN OF MARATHON, IN GREECE, OF WHICH THE HILLS HERE REMINDED HIM.

CRADLE OF WEST TEXAS CATTLE INDUSTRY. AMONG THE FIRST NOTED RANCHERS HERE WERE MAYER M. HALFF AND BROTHER, OWNERS OF THE FAMOUS CIRCLE DOT BRAND. ⊙

ORIGINAL GATEWAY TO THE BIG BEND NATIONAL PARK.

(1965)

Historical marker near Marathon.
— Courtesy Jesse Halff Oppenheimer

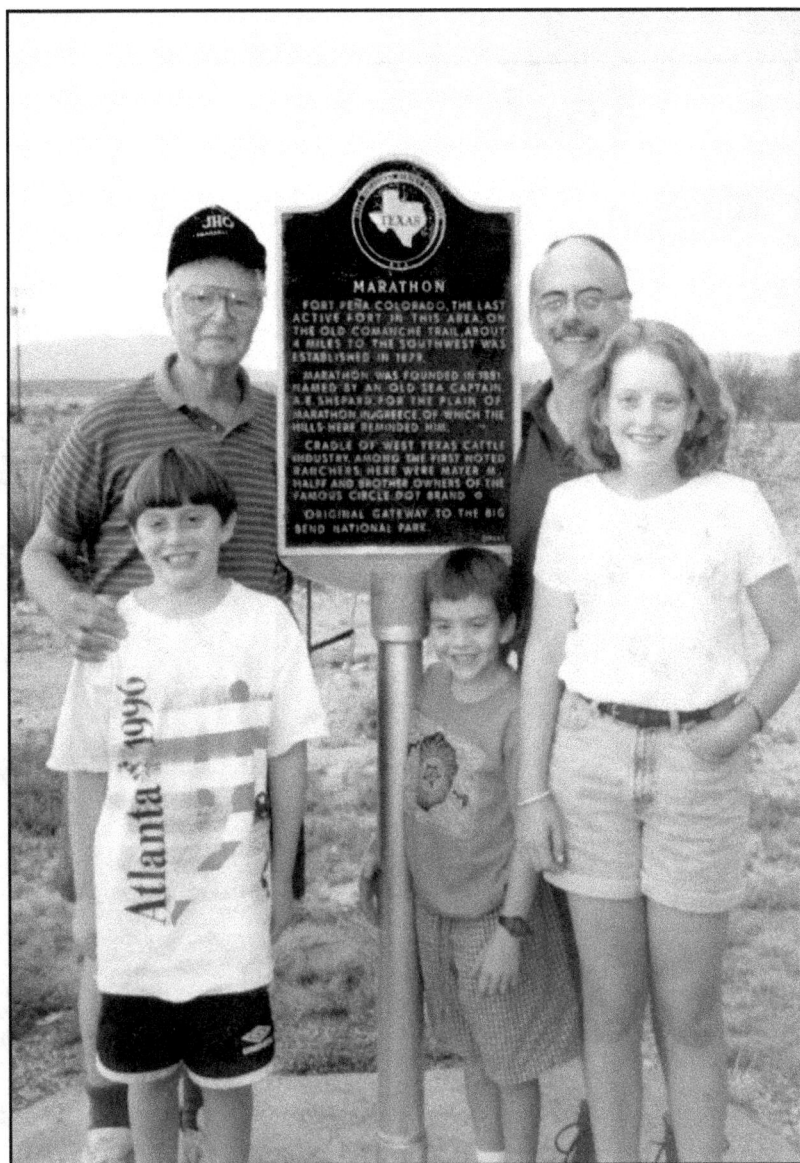

Three generations of Mayer Halff descendants in Old Circle Dot Country:
(back) Jesse Halff Oppenheimer, J. David Oppenheimer;
(front) Daniel Oppenheimer, Jacob Oppenheimer, Rebecca Oppenheimer.
— Courtesy Jesse Halff Oppenheimer

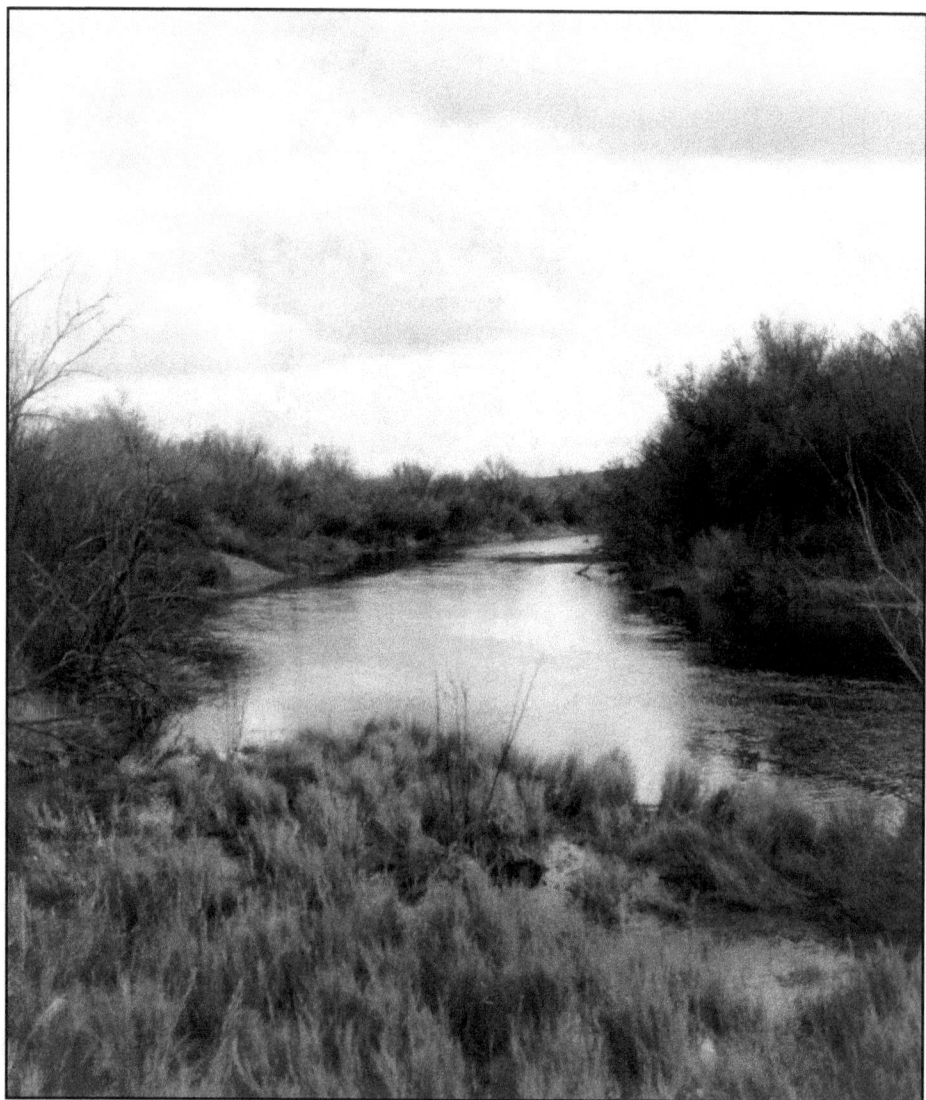

Site of Pontoon Bridge on the Pecos River.

— Photo by author

Narbo's Crossing on the Pecos River as seen in 1999.
— Photo by author

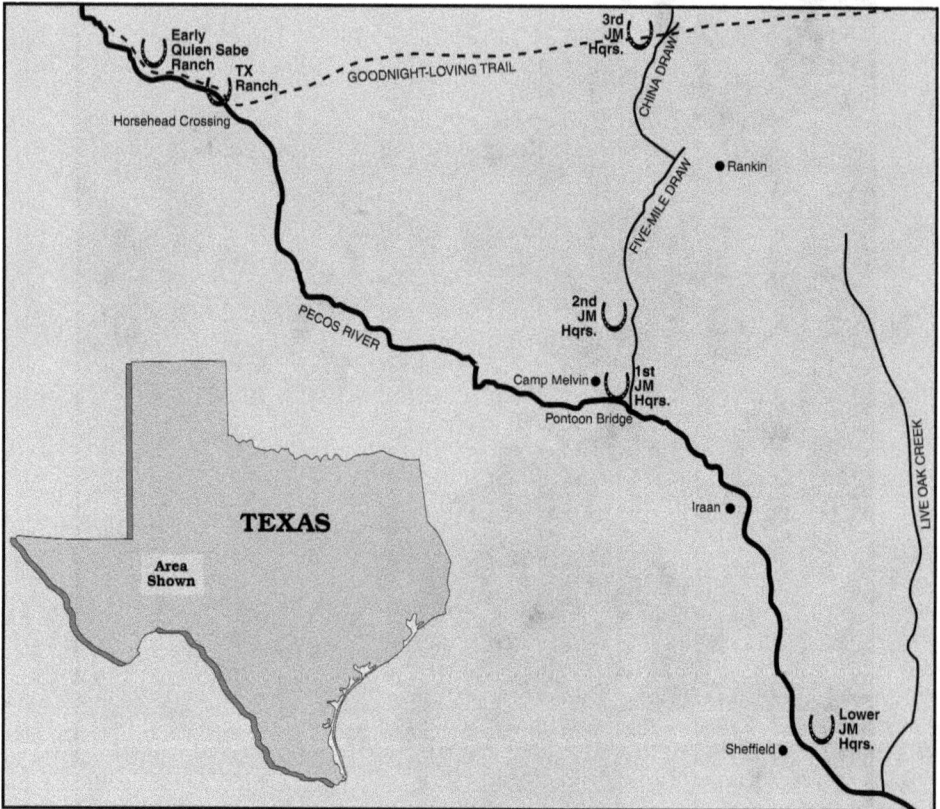

Mayer Halff's JM Ranch. (LMG Designs)

Cowboys on roundup at JM between 1901 and 1906.
— Courtesy Vera Dell Allen

JM cowboys around chuck wagon between 1901 and 1906.
— Courtesy Vera Dell Allen

JM roundup between 1901 and 1906.
— Courtesy Vera Dell Allen

JM cowhands holding remuda in rope corral between 1901 and 1906.
— Courtesy Vera Dell Allen

JM wagon outfit.

Artifacts at site of JM headquarters on Five-Mile Draw in 1999.
— Photo by author

The Elliott house, one-time JM headquarters, as seen in 1998.
— Photo by author

Stone milk shed at the JM-Elliott headquarters.

— Photo by author

Remnants of the rock corral at JM-Elliott headquarters.
— Photo by author

Bunk house at the JM-Elliott headquarters, as seen in 1998.
— Photo by author

*San Angelo saloon at the turn
of the twentieth century.*
— Courtesy Vera Dell Allen

*One-time JM cowhand
O. W. Parker.*
— Courtesy Vera Dell Allen

Ruins of Camp Grierson's Spring.
— Photo by author

The old military road leading down to Grierson's Spring.
— Photo by author

Water at Grierson's Spring.
— Photo by author

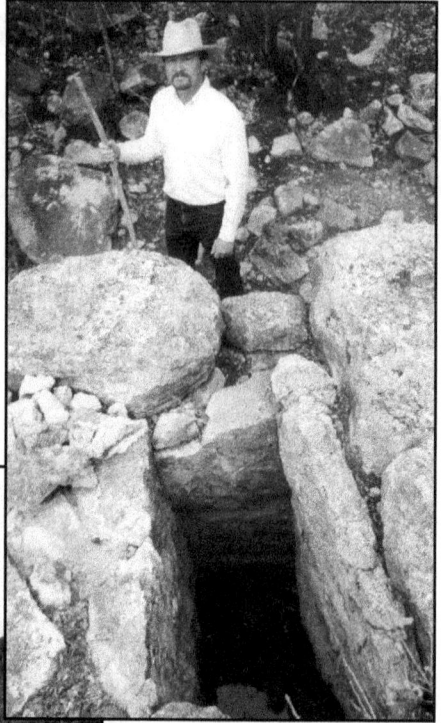

*Landowner Sid Sullenger at the
original spring box at
Grierson's Spring.*
— Photo by author

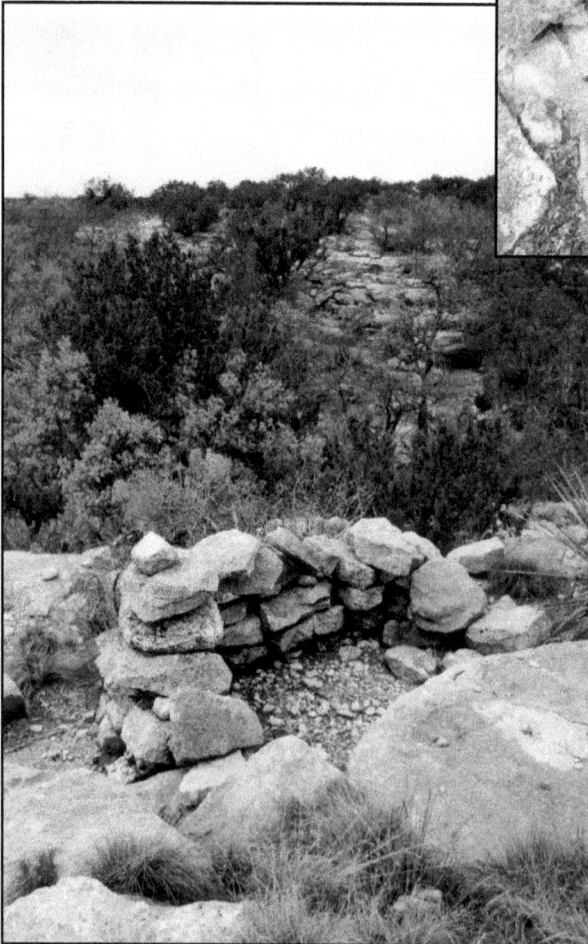

*Military observation
post overlooking
Grierson's Spring.*
— Photo by author

Chapter 9

A Big Bend Brothel

At the height of drought in the spring of 1889, the U.S. Army erected a pig pen on Peña Colorado Creek. Mayer, at the ranch May 21 to ready a cattle shipment to Sabinal, was appalled—the hogs were polluting the water to the detriment of Circle Dot headquarters a short distance downstream. He immediately dashed off a letter to post commander Lieutenant George H. Morgan and politely requested that the army remove the "unsightly" pen and adjacent rubbish.

In his curt reply two days later, Morgan asserted that the hogs did not pollute the water "to a damaging extent" and that the terms of the lease did not require the army to "make the pens ornamental." Morgan, evidently an uncompromising type, furthermore proceeded to make a request of his own. In recent weeks, Mayer's cowhands had driven thousands of cattle through the post en route to the railroad, creating, said Morgan, a "great annoyance from dust resulting from the drought." He demanded that Mayer limit herds to a corridor "across the southeast corner," as specified by the lease. Morgan further asked that Mayer inform him in advance of any such drives.

In his gentlemanly reply on May 22, Mayer said that he had instructed John E. Shoemake, now Circle Dot manager, to adhere to Morgan's instructions regarding movements of herds across post grounds. He also once again called Morgan's attention to the pig pen and the problems it posed for Shoemake's

71

family. (In a letter three years later, Mayer would be far less supportive of Shoemake, instead pondering, "If there only was a way to . . . get rid of Mr. Shoemake and his gang.")

Acknowledging Mayer's letter May 22, Morgan continued to be obstinate regarding the hogs, offering only to move the pen "up the stream to some convenient place at the expense of M. Halff & Bro." However, on May 24, Brigadier General Stanley, of the Department of Texas Headquarters in San Antonio, criticized Morgan for ever allowing the pen in the first place.

"On general principles," wrote Stanley, "the putting [of] a pig pen upon a running stream or a water-hole in a country where water is so scarce as the neighborhood of Peña Colorado, is not the proper thing to do. You will therefore remove the pig pen immediately."

Stanley closed with a bit of advice for Morgan: "It would be just as well to get along with the Halff Brothers without unnecessary trouble."

That same month Mayer leased his Peña Colorado store building to F. H. Oswald, a decision that soon led to more difficulties between Mayer and Lieutenant Morgan. Oswald, a civilian employee of the post, immediately left his job as a wheelwright and converted the structure into a saloon. By May 20 he already had angered Morgan by refusing his order to close for the night. Years later, saloon patron Alfred Gage provided insight into the disreputable nature of the establishment.

"[Gage] said there'd always be a bunch of drunks out there in front, shooting at everything in sight," related James Travis Roberts, whose father, J. J. Roberts, was a wrangler on the Circle Dot. "He said, 'I'd loaf around there till they ran out of ammunition and needed another drink. I knew if I left there [before then], they'd shoot at me just as soon as I run. So I'd wait till they went back inside, and then I'd get my horse and slip around the hill and leave.'"

On the afternoon of June 17, the drunken Oswald wandered outside and did just as Gage described—he opened fire toward nearby cattle and horses. When the sergeant of the guard ordered him to stop, Oswald staggered over to the post administration building and burst in on Lieutenant Morgan. The post commander endured Oswald's cursing, but when the man threatened his life, Morgan confined him to the guard house and notified Mayer.

The problems relating to Camp Peña Colorado only escalated in succeeding months. On July 26 Morgan complained to his superiors that M. Halff & Brother was guilty of trespass, habitually grazing 100 to 3,000 cattle and horses on the military reservation "during the dry weather when our animals need it the most." Moreover, said Morgan, the Halffs housed a saloon which operated in defiance of his wishes regarding policy and hours. "[The establishment] is a nuisance and should be abated," he wrote.

On into fall, the lieutenant remained frustrated by the disciplinary problems the unsavory bar created among his troops. On October 18 he finally tried to force Oswald into adhering to certain military regulations. Citing "General Orders No. 75," Morgan informed Oswald that not only were post traders prohibited from selling hard liquor, they could sell light beer and wine only in "unbroken packages to officers and to canteens." He further stated that regulations permitted the post treasurer to tax Oswald a maximum ten cents per month for every officer and enlisted man at the camp.

In November, however, the army replaced Morgan as commanding officer, and the threatened measures were never imposed on Oswald.

Decades later, Morgan still smarted from his dealings with the saloon during his tenures as post commander in 1886, 1888, and 1889. "Personally I was almost continually in hot water with the [land] owners," he wrote, "on account of my opposition to the saloon they maintained on the premises."

Morgan's successor, Captain Oscar Elting, inherited the saloon issue, doubtless to his chagrin. Mayer, meanwhile, seemingly underestimated the problem, despite Morgan's letters, for Oswald had come highly recommended as a "sober and correct" individual. Perhaps Mayer had dismissed Morgan's reports as the overreactions of a man who, judging by the pig pen matter, had proven himself impatient and intransigent.

Finally, in February 1890, the situation came to a head, and it was Mayer—the upright, highly respected, scandal-free pillar of San Antonio society—who emerged red-faced: He was housing a brothel at Peña Colorado.

"This man Oswald," reported Captain Elting on February 14, "since the night of the 11th inst. [February 11], has been harboring two prostitutes, one white and one black, this to the demoralization of the troops here. . . . I have ordered these women off the premises. . . . During a portion of this time Oswald has been very drunk and noisy, and on the night of the 12th inst. [February 12], he cut a soldier of this command on the neck, very nearly severing the juglar [sic] vein."

Elting laid the blame squarely on M. Halff & Brother, which, he claimed, had violated terms of their lease with the government. The captain specifically cited a clause in which the brothers had granted the army "the quiet, peaceable possession and occupancy of the . . . premises." Elting closed his report to the assistant adjutant general with a recommendation: "I would suggest that if this drinking saloon is to be tolerated, that M. Halff & Bro. be advised to send some decent man here to relieve Oswald."

Upon learning of Oswald's latest transgression, Mayer wasted no time instructing Elting to throw him off the post and temporarily close the saloon. Elting immediately ordered Oswald off the premises, but the man refused to do so as long as he held a lease. Elting, in no mood to joust, proceeded to deliver Oswald an ultimatum—vacate within forty-eight hours or be removed by force. Only then did Oswald leave.

By May 1890 Mayer had found a reputable replacement for Oswald, a man named Dickey who "sells nothing stronger than beer," Elting reported on May 7.

Once such a den of iniquity, the Halff building went on to gain a measure of dignity in succeeding years; it housed a school where the mother of Circle Dot cowboy J. J. Roberts once taught.

While the saloon affair was building to a climax, nature toyed with the drought-stricken Circle Dot and JM. Abundant rains fell in late July 1889, rejuvenating the region enough for an El Paso observer to crow on July 24 that "the grass is green as a wheat field and the cattle are, comparatively speaking, in 'clover.' . . . [They] will be 'hog fat' before winter."

But as the ensuing months would tell, the rain was only a reprieve, not a drought-breaker.

By early December the Pecos country was on the brink of disaster again. On December 3 George B. Loving of El Paso described the Pecos range as "played out as a cattle country. The foothills and adjoining highlands and plains still grow fine grass, but the long, wide valleys next to the river are almost entirely destitute . . . . Cattle through this immediate section [near Pecos City] are not in good condition, and if the winter should be a long, severe one, the loss will be heavy."

An apparently mild winter helped matters, however, and as February 1890 approached, West Texas stockmen prepared to point 40,000 cattle up the trail to Indian Territory. Mayer planned several drives of his own, both from his western range and from his South Texas holdings, and in February he purchased the requisite horses in the lower country. He hoped to have his South Texas cattle on the trail by mid-month and his JM and Circle Dot beeves underway by March 1. "His destination is north," said the *Texas Live Stock Journal*, "and if not sold sooner they are liable to reach the north pole as food for the Esquimaux [Eskimo]."

The reason for Mayer's sudden return to the drive en masse is unclear, but freight rates and swelling competition for cattle cars may have played a role. Simultaneous with increased drives from Texas came a change in quarantine laws for Texas fever. Previously, the Trans-Pecos had been unrestricted, but officials now extended the quarantine line west to the 103rd Meridian. Fortunately for Mayer, the Circle Dot remained unaffected. All the same, the March 1 *Texas Live Stock Journal* called the new regulation a "calamity . . . [for] Western Texas," and pondered whether Trans-Pecos stockmen would be better served by maturing their steers on their home ranges.

Nevertheless, Mayer had several herds on the trail by early spring. Many of his drives originated in Southwest Texas, but the JM likely supplied its share of cattle as well. It was none too soon for his Pecos beeves, now bony and starving. George B. Loving, a passenger on the T&P in late March, described the besieged West Texas area:

"Cattle are very thin, and judging from the number that can be seen from the car window lying dead around the water holes, I conclude that the loss outside the pastures must be pretty heavy. The country between El Paso and the Pecos is . . . very dry and but little grass left. Unless favored with rain soon (which is hardly probable), the loss . . . will be very heavy."

The situation in the Peña Colorado area was only marginally better. Frank Collinson, manager and part-owner of the Coggin & Parks outfit fifty miles south of Marathon, reported his cattle in good shape despite extremely dry conditions.

By the last week in March, one of Mayer's herds, 2,000 head from likely the Wish (or Kelly) pasture near Sabinal, had passed through Uvalde County. Simultaneously, he had another 2,000 cattle northbound from Peña Colorado. Two more Halff & Brother droves, each with 1,900 head, passed through Concho County in April on the way to summer pasturage in Kansas.

Many other Texas herds, meanwhile, continued their march to Indian Territory. Moreover, the T&P was doing a booming business rushing drought-stricken West Texas cattle to points north of the Red River. In the first two weeks of April, 500 car loads shipped out from Midland, Colorado City, and Abilene, and the railroad already had orders for 1,500 more cars. Rail destinations for the estimated 75,000 to 100,000 cattle were the Creek, Osage, and Otoe nations, the latter two of which had very limited grazing lands.

But suddenly, the Texas cattle industry faced a crisis that originated in the highest office in the land: President Benjamin Harrison handed down an order prohibiting cattle from entering Indian Territory. Furthermore, he gave stockmen only until October 1 to remove existing herds, a deadline he later extended to December 1.

Texas cowmen had already targeted tens of thousands of cattle for the Territory, many of them in the form of droves even now on the march. It was a troubling development, prompting the *Cheyenne Live Stock Journal* to editorialize that the question of the cattle's fate was a matter of "grave importance."

Meanwhile, range conditions along the Pecos were growing critical after eighteen months without productive rain. Winfield Scott of Colorado City reported that all outlying water holes were dry and that it was fifteen miles from the river to a "spear of grass." One party, related Scott, counted 4,000 cattle carcasses in the Pecos over a forty-mile stretch. Thousands more dotted the flood plain and beyond. On the JM side, especially, even the grass roots had been trampled to dust, courtesy of three calamities: overstocking, the Big Drift, and now a hellish thirst for three and a half of the last five years. Lan Franks, writing eleven years later, fittingly referred to this period as "the most disastrous years ever known in the cattle industry."

The Burro and Organ mountains in New Mexico, grazing lands for thousands of Mayer's cattle, were almost as hard-hit as the Pecos. Wrenched by drought since December 1889, southern New Mexico had turned "bare and brown," reported a Silver City observer. "Cattle are dying by the hundreds daily. . . . The grass has been eaten down close to the ground in the vicinity of water courses, and herds go miles into the foothills for feed, where they remain till driven in by thirst. . . . They drink their fill and lie down, never to get up. Thousands of carcasses of dead cattle lie rotting in the sun."

In early May showers finally blanketed West Texas, and follow-up rains during the summer turned wasteland into grass land. In early August, cattleman James D. Kennedy already boasted of fat cattle in the Pecos country. But it was not until soaking rains favored the region in September that cattleman finally considered the two-year drought at an end. The Pecos valley was "blooming like a rose," said the *Texas Live Stock Journal*, with grass as ample as ever on the JM range and others.

"The large herds that had for years been compelled to leave the Pecos valley and rustle for grass in the foot hills and on the plains have now returned to their old ranges in the valley," noted the *Journal*. "The Pecos country once more presents a prosperous appearance."

But, as always on the Pecos, prosperity came at a price for cattlemen. As the *Pecos Valley Register* reminded its readers that fall, the thousands of carcasses rotting along the river bespoke a harsh reality in a harsher land—"many die, that those which survive may prosper."

With the new restriction against pasturing in Indian Territory, the majority of Mayer's drives in 1890 may have ended in Kansas, the ultimate destination of many aimless herds that year. Meanwhile, the relocation of existing Cherokee Strip herds—possibly including Halff beeves—got underway May 15 with a west-to-east roundup. By early July most of the cattle were ready to be sorted and driven out of the Territory. With quarantine laws making it unprofitable to relocate the herds to northern ranges—and high tariffs making it equally counterproductive to send them south to Coahuila, Mexico—cattlemen had little alternative but to flood the market and risk sending prices on a downward spiral.

However, some stockmen considered this "final round-up" to be more than a dollars-and-cents matter.

"They seem to regard it as the last sad rites to be performed over their departed occupation," reported the *Kansas City Journal*. "They want to remember it [the Territory] as it was, and they are on the ground to see the end. . . . The great plains and river valleys that have pastured as many as 2,000,000 head of cattle at a time, will be left to the government and the Indians . . . , and the great cattle ranges of the continent will be almost exclusively confined to the Rio Grande and Pecos river valleys."

By the December 1 deadline, the Cherokee Strip Live Stock Association, in which Mayer likely was a stockholder, had purged its vast range of cattle, two years before its lease with the Cherokee Nation expired. The association contemplated suing the federal government for one million dollars in damages.

But Indian Territory would rise again as a cattle country, just as the Pecos had already done in the past few months. The latter range had greened due to the very thing that had bared it, for the long drought had nurtured as well as devastated. Through losses and the forced relocation of cattle, Pecos herds had been depleted by as much as fifty percent. Now, for the first time in years, the range was no longer overstocked. By early December, beeves along the river were in such excellent condition that stockmen expected few winter losses.

Meanwhile, the feared downturn in cattle prices stemming from the Indian Nations crisis had become a reality. The previous July, Mayer had netted $2.60 per hundred weight for steers in Kansas City and $1.87 for cows in Chicago, but by early December, market prices had plummeted to a fifteen-year-low. It wasn't an encouraging start to the Gay Nineties, but better times—and worse—lay ahead for Mayer and his fellow cattlemen.

Chapter 10

Paradise on the Pecos

In late summer 1890, Mayer took a step that eventually led to a Halff kingdom rising out of the deeded JM. Within a few years, only a single ranch—the old Block outfit north of present Rankin—would keep the Halff empire from stretching uninterrupted across an incredible ninety-three miles from the mouth of Live Oak Creek, up the Pecos to Pontoon, north to Midland, and then east to within ten miles of Garden City.

Evidently to no avail back in 1886, Mayer had tried to lease university land up Five-Mile Draw from his Camp Melvin river section. In the following years he had continued to see the potential in the property, believing it only a few windmills away from being prime grazing country. JM Draw already had occasional depressions that filled with runoff during rainy periods. Meanwhile, the state had rethought its policies, setting aside the property in 1887 and 1889 for "the benefit of public free schools, the university, and asylums." The result was that, on August 20, 1890, Mayer came to terms with the state for a ten-year lease of thirty-eight sections, or 24,320 acres, at an annual rate of three cents per acre, a mere $729. Mayer—ever the wise range manager—could now ease the burden on his river sections without stifling plans for increasing his herd and even maturing his steers.

He presumably began by fencing off a large pasture, almost

certainly untrampled university land, during the unusually mild winter of 1890–91. In years past he had transported his young JM steers exclusively to northern regions for maturing, but now, ambitiously, he hoped to fatten them on their home range.

His steer crops on other ranges were another matter. Much to his relief, no doubt, Indian Territory was more than just a memory in 1891. Despite the eulogies delivered over it by cattle-men the year before, the Territory again came alive with Texas beeves, courtesy of the federal government. With the dawning of spring, Mayer played a role in its rebirth, stocking the region around Catoosa (just east of present Tulsa, Oklahoma) with more than a thousand beeves ripe for maturing.

Early May of 1891 brought ten days of abundant rain to the Pecos, a development which, with a rise in cattle prices, left cat-tlemen exuberant. "They have been paying the fiddler for a long time," reflected one Pecos City resident on May 12, "but think now their time has come to dance." Simultaneously, the U.S. Department of Agriculture announced that ticks had been isolated as the cause of Texas fever in cattle. Although cattlemen initially met the report with less than unanimous acceptance, they eventually would wage war on the parasites. Meanwhile, the Crockett County general roundup was underway on the Pecos, and out of it would come choice, young steers that Mayer would assign to his new JM pasture.

With 35 cowboys, 200 horses, and a chuck wagon, the roundup commenced May 1 on the H-Bar outfit and worked upstream through the JM range. G. A. Noble, who was now JM foreman, bossed the operation, which was to conclude at JM headquarters at Pontoon Bridge. From there, JM hands were to cut out selected JM steers and turn them into Mayer's new pas-ture, where they could mature for the summer and fall market.

As branding fires burned hot along the river and cowboys scoured the deep side-canyons for strays, Mayer broadened his vision yet again. On May 18, M. Halff & Brother purchased the South Texas ranch of Edward Prince for $20,301 plus four notes (the last due in November 1892) totaling $56,315. The 18,516-acre "Prince Pasture," as Mayer would call it, comprised twenty-two sections in McMullen County, four sections in LaSalle County, two sections in Atascosa County, and 596 acres in the "four-corners" region of Frio, LaSalle, McMullen, and Atascosa counties. Three days after his initial purchase, Mayer acquired an additional 3,200 acres in McMullen from Virginia M. Prince. In its new form, the Prince Pasture was sprawling, scenic, and filled with promise, and Mayer wasted no time in stocking it with cattle.

For tax purposes in 1891, Mayer declared 3,500 cattle and 100 horses in McMullen, where he had maintained a presence for several years. Fifteen hundred of the beeves ranged on leased acreage, but the other 2,000 already roamed the new Prince Pasture. The appraised worth of the two herds was $19,000, while the horses were taxed at $15 apiece.

In an mid-June interview with the *Texas Live Stock Journal* during a Fort Worth visit, Mayer made reference to his sheep interests. Although back in September 1882 he and Solomon had purchased 2,299 sheep, 43 bucks, and 441 goats for $5,735 from L. Marulanda & Company in Webb County, he had remained essentially a cattleman. In fact, he eventually had disposed of the flock through outright gift or by selling at a loss (his son Henry, writing in 1903, used the term "gave away"). By the time of his remarks to the *Journal*, Mayer's involvement in the sheep industry was not as a raiser, but as a wool wholesaler with a vested interest in numerous Texas flocks.

Even during their partnership with Abraham Levy, Mayer and Solomon had traded in wool, and by the late 1870s the brothers were hailed as "wool kings" in San Antonio, a city known as "the wool market of the world." M. Halff & Brother regularly advanced monies to sheep raisers, and Solomon, in his role as manager of the wholesale operation, maintained a high profile in the Texas Wool-Growers Association. Its members named him to the board of directors in June 1890 and to the finance committee a year later. Respected for his insight, Solomon was outspoken on wool matters, even making a spirited defense of the honesty of Texas sheep raisers at the 1891 meeting. When the association president responded by telling attending sheepmen that Solomon was entitled to their thanks, Solomon demurred. "Mr. Halff," reported an observer, "said he considered it no compliment to tell an honest man that he wasn't dishonest."

Two-hundred-fifty-pound Solomon, although less physically active than Mayer, nevertheless found plenty to keep him busy. In addition to his duties with M. Halff & Brother and the Texas Wool-Growers Association, he had become involved with a new financial institution in San Antonio. On November 20, 1890, he had attended the organizational meeting of Alamo National Bank, and two weeks later he had become its initial vice president. He and Mayer each owned fifty shares, or two percent, in the bank, which was chartered on March 2, 1891, with $250,000 in capital stock. Stockholders included Charles Hugo, who was chosen as president; J. N. Brown, cashier; P. H. Swearingen of Brenham; William Huerman; Ernest Steves; G. Schmeltzer; H. Elmendorff; and Charles Schreiner of Kerrville, who later would play a prominent role in Mayer's ranching endeavors.

In the summer of 1891, Mayer continued working toward improving his JM herd. In late June he dispatched JM manager G. A. Noble to a Hereford farm near San Angelo to inspect native bulls bred by John Harris. Noble was so impressed with the one- and two-year-old Herefords, which subsisted entirely on grass, that he purchased thirty-two bulls for the JM. While his cowboys started the animals for the ranch, which lay several days' drive to the southwest, Noble went up the North Concho River to close a trade with W. B. Hiler for 300 of the region's finest cattle.

September 2 marked an important milestone in fifty-five-year-old Mayer's personal life, for he and Rachel had now been married twenty-five years. To celebrate their silver anniversary, Mayer took his wife, who was almost forty-six, on a trip to Saratoga, New York. They presumably returned in time for Mayer to oversee fall cattle shipments to the Chicago and Kansas City markets. His sale of 1,818 beeves (mostly from Indian Territory and Elgin, Kansas) for $34,402 bespeaks the astounding success of a cattleman operating at a time when a cowboy gladly rode for a dollar a day.

Over the next dozen years, Mayer would be no stranger to Elgin, a town situated in Chautauqua County on the Kansas–Indian Territory line. During the 1890s the small rail station gained a reputation as the world's busiest shipping center, serv-

ing cattle operations in both Kansas and Indian Territory. The
exact location of Mayer's range is uncertain, but Chautauqua
County's fertile Caney River Valley and rugged oak hills offered
prime grazing.

While Mayer was shipping in the fall of 1891, trouble developed on his unfenced grazing land near the Burro Mountains in
New Mexico. A couple of area outfits discovered that their
brands had been "run" (changed with a running iron) into a
brand claimed by a man named Langford, who apparently managed Mayer's operation. When a house on the property soon
burned under mysterious circumstances, Langford informed
Mayer, but failed to mention the branding controversy. Instead,
Langford suggested that vengeful cattlemen had torched the
structure because the ranch had denied free-ranging strays the
right to water.

Mayer, obviously distrustful of Langford's account, quickly
wrote W. W. Cox, an old friend who ranched at Cook's Peak
north of nearby Deming, New Mexico, and asked him to inves-
tigate. Mayer's friendship with Cox, to whom he granted considerable authority, was improbable, to say the least. Mayer had
welcomed Cox into a small circle of highly trusted lieutenants
despite Cox's 1878 conviction on murder charges and his role
in a long and bloody Texas feud.

Thirty-seven years old, Cox was the son of James Cox, a
principal in the so-called Taylor-Sutton feud which eventually
spanned thirty years and two generations. In approximately
1870 the elder Cox joined the Regulators, a law enforcement
group sanctioned by the Texas Legislature and charged with
bringing in the Taylor clan for the 1867 murders of two U.S.
Army soldiers. But the Regulators walked a thin line between
being peace officers and outlaws, and their search for justice
evolved into the Taylor-Sutton feud. When Regulator Cox
learned in 1873 that he was a suspect in a murder, he started for
Helena in southeast Texas to answer the charges, only to be
ambushed and killed. No arrests were made, but the younger
Cox and the Sutton faction believed Dr. Phillip Brassell and his
son George Brassell were responsible.

At only nineteen years of age, W. W. Cox took up the fight,
and when the Brassells were killed in 1876 a grand jury indicted

Cox and six other Sutton sympathizers for first-degree murder. Facing the death penalty after his conviction on April 17, 1878, Cox was spared only after a couple of bizarre twists. The court of appeals in San Antonio upheld the verdict, but a clerk lost the judgment, forcing the court to reopen the case. Cox then won a reversal and a new trial on a technicality that almost strains credulity: The indictment, which should have closed with the words "against the *peace* and dignity of the *state*," actually ended with "against the *piece* and dignity of the *statute*."

Posting bail and clinging to freedom on the basis of two misspelled words, Cox hired on with Mayer in the Big Bend while awaiting a new trial. He would have a long wait—twenty-three years, in fact. Meanwhile, he gained Mayer's trust and friendship, possibly assisting him in locating grazing lands in New Mexico before venturing to the territory on his behalf. In 1893 Mayer would finance Cox's purchase of a ranch near San Augustine Pass in the Organs and provide him with sheep—perhaps the Webb County flock Mayer had secured in 1882.

There were yet other indications of mutual respect between the two men: Cox named his eldest son Halff Riley after Mayer, and when Cox's new trial in the Brassell case finally was set, Mayer was one of seven prominent stockmen or bankers who posted his $10,000 bond. Mayer's support of Cox was evidently well-founded; the district court in Seguin ultimately dismissed the case against him.

But in 1891 Cox was still living under a cloud. Despite this, Mayer entrusted him with ferreting out the facts about the house-burning. On December 24 Cox reported to Mayer by letter and informed him of Langford's apparent skill with a running iron.

"There is no animosity nor ill will towards you [by area cattlemen] in any sense, and if this house was burned purposely, it was on Mr. Langford's account," Cox noted. "Such a stink has been raised about these burned stock, Langford refuses to claim them. . . . That he *thought the fire was caused on account of the water being cut off* is too thin. The water never was cut off from stray stock."

Cox added that he intended to confront Langford about the matter on December 26 and replace him, ideally with a Mr.

Kersey. "I will take a man with me when I go," he wrote, "so if he [Langford] wants to vacate right away I will put the other man in charge." Obviously, Mayer had authorized Cox to act decisively.

Meanwhile, the Pecos was up to its old tricks back at the JM. If there was anything Mayer had learned about this moody river, it was to be prepared for its relentless drought/rain, drought/ rain cycle. Now, after fifteen months of prosperity, the Pecos was ready in late December to deal misery yet again.

As always in this arid country, the latest drought would go unrecognized for months; the difference between an ordinary dry spell and a damning pestilence was only a matter of duration. By late the next summer, however, the *Texas Live Stock Journal* would describe cattle along the river above Pecos City as nothing but "living skeletons."

One of the things Mayer had on his mind this season was a new regulation that still left the JM under quarantine for Texas fever. Mayer's ranch was one of the uppermost ranges on the Pecos still under restriction; the quarantine line now struck the river at the northwest corner of Crockett County, only twenty-three miles above Pontoon Bridge. While freeing every upstream range (such as the old TX country) from quarantine, the ruling prohibited the JM from selling to buyers in northern regions such as Colorado, Wyoming, Montana, and the Dakotas. The ruling had no effect on the JM's right to ship to market, however, only against mingling its potential "carrier" cattle with northern range animals susceptible to the deadly fever.

The Santa Fe Railroad had reached San Angelo in 1888, and by 1892 the JM looked to that city as its primary shipping and supply point. Mayer was in San Angelo in mid-March handling a shipment to Indian Territory, and three days later JM manager G. A. Noble arrived to secure rail cars. The arrangements made by both men likely concerned the same herd of 1,800 steers already on the trail from the Pecos. The animals probably were shipped to the Cherokee Strip, where Mayer was grazing 1,500 head in mid-July by the authority of Jake Guthrie, a Shawnee and adopted citizen of the Cherokee Nation.

The Strip was filled with cattle by that time, for Texas shipments to the Territory in the spring of 1892 were the largest ever. Not only did stockmen consciously reduce cattle numbers on their Texas ranges, but the Territory was now the only outlet for animals originating below the quarantine line. The April 23 *Texas Live Stock Journal* assessed the matter bluntly: "Indian Territory . . . must therefore be the dumping ground for fully 200,000 Texas cattle during the present spring." A week later, the *Journal* increased its estimate to 300,000 to 400,000 head.

By the end of May, the rush to Indian Territory was over and stockmen like Mayer already were reaping benefits: cattle throughout the Territory were taking on flesh quickly, courtesy of plentiful rains and lush grasses.

The depletion of cattle on Texas range land was fortuitous timing for outfits in the Trans-Pecos, where the drought had taken a stranglehold by late May. The JM and other ranches east of the Pecos had been buoyed by recent showers, but cattle losses were imminent on the Circle Dot and particularly on Mayer's New Mexico grazing lands.

"We have not had a particle of rain yet, and apparently no prospects of any," reported W. W. Cox from Cook's Peak on June 17. "I have just returned from Las Cruces . . . and found that country, if possible, looking worse than this."

During Cox's Las Cruces trip, he inspected Mayer's properties in the Organs. He found the situation at Dripping Springs, north of Rattlesnake Ridge, especially frustrating. A man named George Beasley had located at the springs three years before and refused to leave. Cox tried to evict him yet again, and when the nester failed to vacate, Cox entered suit against him for six hundred dollars for three years' use of water.

At another of Mayer's Organ tracts, which Cox referred to only as "Moran," he found things more orderly; employees had built a small wooden house and had worked eighteen days on a well. Seven miles northeast of San Augustine Pass, Cox also investigated Alamo Springs, near which two miners had staked claims. Six years later, Mayer would sell Cox the Alamo tract, as well as Ruby Mining Claim on the northeast slope of adjacent Black Mountain.

By early July the showers of spring were a distant memory to Pecos River outfits, which now lost cattle daily to starvation. One prominent stockman reported that he already had lost

1,500 head, and that the toll on some herds was even greater. Range conditions were little different in the Big Bend, where Mayer had recently entered into partnership with H. H. Carmichael, who had sold him a ranch in Bandera County five and a half years before. Sometime before the summer of 1892, the two men leased a pasture in the Alpine area from George F. Crosson Sr. and stocked it with cattle. As forage diminished and nourishing rains failed to come, they had no alternative but to seek other pasturage.

In summer they leased a pasture about ten miles south of Alpine and threw five hundred cattle on it from the Crosson range. More beeves remained behind, too weak for rounding up or driving. By early August the critical forage situation had not improved for the new partners. "The pasture they moved to is very dry—no rain in it this year and has had sheep in it last winter," Frank Collinson wrote to S. R. Coggin. "So the prospect is not very good for them. . . . The whole country to San Antonio is in the same fix and north to Colorado, and all through Mexico."

In late July, scattered showers had peppered the Pecos—the first rain in seven months along certain stretches—but the rescue of the region's cattle industry demanded a widespread gully-washer. "Stock are still dying," reported a Pecos City observer on August 2, "and unless it should rain soon, some of the herds will be cut down one-half."

Local showers fell across the Big Bend a week later, raising stockmen's spirits, but by mid-August the "old cry of no rain and short grass" prevailed from the Pecos to the Davis Mountains, reported the *Texas Farm and Live Stock Journal*. Cattle losses already had reached fifty percent in many herds, and the calf crop was the lowest in years. For many outfits the only hope in this valley of dry bones was a lingering season of drenching rains.

In late August, they got it, as drought-breaking downpours finally muddied the entire region. By September 6 a Pecos City observer would even lament that "rain is becoming monotonous . . . The ground throughout the Pecos Valley is thoroughly wet and . . . there was some damage done to adobe houses." In another week, West Texas was a sea of lush grasses—the best in seven years, said stockmen—and cattle had begun to flesh out.

The eight-month drought was not the only crisis Mayer

faced in 1892, for trouble was afoot again in Indian Territory. On July 2, C. J. Harris, Cherokee Nation president, dispatched treasurer E. E. Starr to the Cherokee Strip to tally and tax all cattle, including Mayer's herd of 1,500. When the cattle outfits refused to pay, the stage was set for a showdown.

"I would respectfully and urgently request," Starr wrote Harris July 20, "that you take immediate steps to have all United States citizens with their stock of whatever description (except those who have paid their tax) removed from the Cherokee lands west of 96 degrees."

Starr's report evidently triggered an industry-wide nightmare for stockmen, for by early August several troops of U.S. Army cavalry had orders to evict all intruding cattle from the Cherokee Strip. With quarantine laws preventing cattlemen from driving through Kansas, many outfits had little choice but to market their cattle at prevailing prices. Mayer himself shipped cattle from Catoosa, Blackstone, and Muskogee, as well as from his Kansas holdings near Elgin, Hamilton, and Protection.

As if drought, eviction, and low cattle prices weren't enough trouble, Mayer also had to deal with horse theft in Southwest Texas during the latter part of the year. By separate letters in late summer, the Edwards and Uvalde county sheriffs notified him that Abner Alexander, who may have assumed the name, had sold four head of horses whose brands had been run from the Halff Brothers' H–half-H to an HA. Purchaser was White and Littlefield, who had shipped two or three of the animals to Indian Territory but had kept a roan at their ranch fifteen miles south of Mason.

Both lawmen, Ira L. Wheat of Edwards County and H. W. Baylor of Uvalde County, were convinced that Alexander had stolen the animals in Kinney County. Wheat urged Mayer to arrange for the Territory horses to be shipped back to Texas for identification. "You would be the cause of convicting one of the hardest men [Alexander] in Texas," noted Wheat, "although he is only 21 or 22 years old."

In early September, Mayer requested A. L. Casparis, who worked for White and Littlefield in Indian Territory or Kansas, to inspect two of the horses in question. Casparis' September 10 response, directed to Mayer in Elgin, Kansas, verified that both animals carried an HA on the left shoulder. He added, however, that White and Littlefield would demand lawful proof of Mayer's ownership before relinquishing the horses.

The situation dragged on into late December without resolution. Undoubtedly frustrated, Mayer wrote to ask Mason County official A. W. Koock to try to recover the roan still in White and Littlefield's possession. When Koock presented J. W. White, part owner of the outfit, with Mayer's affidavit of ownership, White's anger flared and he dashed off an uncompromising letter to M. Halff & Brother.

"I will not give up a horse on as many affidavits as you can stack up in Mason County," White wrote. "If he is your horse you can sue me for him and I will make the man I bought him from a party to the suit. In that way if you get a judgement against me for the horse, I will get a judgement against the man I bought him from. You tell Koock to indict me. I wish you would and I will make it hot for you for false indictment. Try me. You can['t] bluff me. I have got as much money to spend as you have."

Although the outcome of the matter is lost to history, the episode does illustrate that not all cattlemen shared Mayer's concern for integrity.

While Mayer was occupied with thievery and other matters, the U.S. Army was busy vacating Camp Peña Colorado. With Indian wars a thing of the past, the post had outlived its usefulness. Lieutenant Colonel B. G. Dandy gave Mayer the required ninety days' notice by letter on December 1, and two days later military authorities issued the official order to abandon. By early February the reservation had reverted to range land and the buildings were open for Circle Dot use.

The winter of 1892–93 was severe in parts of West Texas, where temperatures plunged below zero and lingered for an unprecedented period. Mexican sheep herders reportedly perished from exposure, and numerous livestock froze standing. But by February 20, when Mayer attended a Southwest Texas Stockmen's meeting in Beeville, the Pecos country was "once more a paradise for cattlemen," said W. D. Johnson, co-owner of the W Ranch in present Winkler, Ward, and Loving counties.

Johnson was not alone in his assessment.

"In the Pecos valley, where a few years ago jack rabbits and prairie dogs had to hustle mighty lively for a living, is now as fine a range country as there is anywhere," cattleman W. R. Curtis observed in late February.

But nothing was forever on the inscrutable Pecos, where paradise was only six inches away from paradise lost.

Chapter 11

Death from the Sky

By early spring in 1893, the Pecos country's outlying water holes were dry and cracked. Consequently, cattle had no choice but to drink from the river, which was heavily laced with brine and other impurities that season. At Horsehead Crossing, Panhandle-bound herds already had suffered such serious losses trying to ford that some trail bosses had pointed their surviving cattle back to their home ranges.

One herd that did persevere up the river early that spring was a Halff & Brother drove of 2,600 two-year-old steers, all of them a little thin after the hard winter. Mayer already had a buyer waiting in Colorado or Wyoming, and now he needed to deliver. Fifty-seven now and busy juggling a wide array of far-flung ventures, he went about other business as the herd dusted the alkali flood plain on up into New Mexico. A few other droves also stayed on the march, spaced ten to fifty miles apart. Mayer's herd reached Fort Sumner without incident, but on a barren flat just beyond, a dark cloud descended, and hard, white rocks suddenly began to rain down with wicked force.

"They said it was *fierce!*" related Halff cowhand Bob Beverly, who learned of the incident from several of the drovers. "They just had to get down and turn the horses loose and put the saddles over their heads."

It was a technique every cowboy learned as a greenhorn, but

this was a hail storm unparalleled in the annals of trail driving. The drumming was deafening as thousands of stones slammed into turf and hopped like a swarm of demons. As the drovers cowered, taking a frightful beating despite their protection, steers and horses dropped left and right, their skulls fractured. When it was finally over, the cowboys, wincing from their bruises, crawled out from under their battered saddles to find themselves in the middle of an astonishing scene—in every direction sprawled hundreds of carcasses, stark against a plain white with ice.

The drovers gathered themselves and waded through, tallying an incredible toll of six hundred steers and almost every horse in the remuda. No other drove in the area experienced anything more than rain.

Cast adrift on the ruthless Pecos with 2,000 steers and virtually no horses, the drovers had no choice but to hold the herd and start a rider back down the long trail to Texas. Mayer, evidently notified by wire, traveled by train to Toyah and met the messenger, who doubtless was exhausted after his 300-mile ride. After hearing a story that beggared description, Mayer was stunned.

"Was there no more cattle in this country but Halff's herd?" he asked.

"Yes," said the drover, "four, five more herds along there on the trail pretty close together."

"And the hail didn't kill any cows for them?"

"No, it didn't hit them, this hail storm didn't."

Mayer shook his head. "Well, I can't understand it. What's the matter with God Almighty? Does he kill the old Jew's cattle and not bother the Gentiles'?"

Accompanying the cowhand back up the hard trail to Sumner, Mayer outfitted the drovers with horses he must have purchased locally. He probably continued north with the herd until reaching a convenient railroad station and taking a train back to San Antonio. No doubt, he was relieved that the situation was in hand. In late spring, however, as the herd pushed up through Colorado, the trail boss wired him of yet another disaster—a blizzard. Soon after hearing the news, Mayer walked inside a San Antonio hotel on business and quickly drew attention.

"Why, it's summertime here," someone told him. "What are you doing all wrapped up with an overcoat on?"

"My God!" he exclaimed. "I got a herd of cattle in a snow-storm freezing to death in Colorado!"

The two calamities on the Pecos-Colorado trace in '93 were enough for Mayer; he never sent another herd north by way of it. Interestingly, his weather-related misfortune may have made him particularly sympathetic to the plight of the wife and six children of cattleman Matt Owens of Cisco that year. A tornado had ripped through their home, killing Owens, a son, and a daughter, and crippling the seven survivors. Nothing remained of their house or belongings except three feet of rock chimney, stone steps, and a single chair. In late May or early June, Mayer donated ten dollars (ten days' pay for a cowboy) to a fund to benefit the family.

Other cattle affairs also occupied his time in the early months of 1893. In March H. H. Carmichael sold five hundred Halff and Carmichael steers to John R. Holland. The beeves ranged in age from one to four years and brought $7 to $15 a head. Moreover, Indian Territory was open again, and Mayer was in Colorado City in mid-March to arrange the shipment of 2,000 cows to the region. He personally received the animals in early April at his Otoe reservation pastures, where he planned to fatten them for market. The cows may have composed only part of his reservation herd; on April 28 the *Texas Live Stock Journal* reported that he had "several thousand cattle on grass" in the Territory.

Meanwhile, another full-fledged drought had gripped the Pecos, where cattle losses on some ranches were mounting. JM animals fared better, however, according to Mayer's mid-June assessment; perhaps careful range management and the Five-Mile Draw pasturage made the difference. He was stocking the latter university land heavily, declaring 1,500 cattle and 30 horses for Upton County tax purposes that year.

He also had acquired 160 acres west of San Antonio in Medina County, where his holdings would grow to 595 acres within six years and to 1,911 acres by 1901. The purpose behind the acquisitions is unclear, as Medina officials never taxed him for a single head of livestock.

Summer rains brought relief to the Peña Colorado area in 1893, lifting cowmen's spirits. Even the Pecos, despite little rain and significant cattle losses, had prospects by July of the best calf crop in three years. Part of the reason was that the river country was not overstocked for a change. Mass shipments to Indian Territory, said cattleman W. P. Birschfield, had reduced Pecos herds to only one-third their 1890 size.

Mayer inspected his New Mexico grazing lands in mid-November and may have returned by way of Peña Colorado, which probably was part of a broad region powdered with a ten-inch snow before Thanksgiving. The precipitation was promising enough for the *Texas Live Stock Journal* to predict "early and abundant" spring grass. However, the succeeding months would prove the *Journal* guilty of wishful thinking.

In late fall, dry conditions on the South Plains and Edwards Plateau set the stage for added trouble on the Pecos, whose grasses evidently had benefited from the same snowfall. Day after day, Pecos-bound herds from the two regions raised a dust over Midland, nurturing a concern by cattleman S. E. Townsend that the river would be so badly overrun that great suffering and loss were a certainty.

Late fall 1893 also brought a disturbing development at the Circle Dot. At a Kansas City meeting, representatives of several state sanitary boards placed Buchel County, which then included Peña Colorado, under quarantine for Texas fever. Mayer reacted strongly, claiming that he had delivered Circle Dot cattle to points throughout the Northwest for years without ever conveying the disease. Even more alarming, he said, South Texas cattle that undoubtedly carried the fever could now enter Buchel, threatening his own susceptible animals.

Mayer was particularly upset in light of the fact that, for years, quarantine officials may have unfairly targeted the deeded JM. Cowman Fred Wilkins noted in early December that in a decade of dispatching herds from Crockett County (where the deeded JM was located), he had never heard of a Crockett animal transmitting the fever.

The skies stayed barren over West Texas on into the spring of 1894, portending imminent disaster for cattlemen. In March, Mayer sent out a drove of 2,000 Circle Dot steers by way of San

Angelo, hoping to find better range in Indian Territory. Solomon, who usually deferred to his brother on cattle affairs, opposed the move. "Better let 'em stay where they are and save funeral expenses, which are less in Texas than in the Territory," he reportedly told Mayer.

By May cattlemen were in dire straits across a vast, parched region and especially on the Pecos, where outfits were paying the price for overzealous stocking during winter.

"The condition of range stock . . . is deplorable throughout Western Texas," reported the *Texas Live Stock and Farm Journal* on May 11. "Cattle have been subsisting on brush and prickly pear. The ground is perfectly bare, no sign of grass roots remaining, and the late rains have afforded no immediate relief except to put water in dry ponds and lakes. . . . Cattle are dying and many of them are too poor to reach the shipping pens for movement to the Indian Territory."

Lan Franks, writing in 1901, described the scene graphically: "In the canyons, along the creeks and around the water holes, the ground was covered with carcasses and the air was burdened with stench. . . . In all directions, the earth was dotted with shining objects: 'twas the sunshine dancing on skinned carcasses."

The losses were staggering. A prominent Tom Green County stockman lost more than 600 three-year-old steers out of a herd of 700. A plains cattle company rounded up only 500 beeves from a pasture which had held 3,000. Another outfit could tally only 500 cattle from a herd 7,000-strong. The cost to the JM and Circle Dot is not known, but few cattlemen survived the ordeal with more than fifty percent of their herds.

Finally, after fifteen hard months, this latest withering plague began to relent. On June 3, 1894, the first rain since August 6, 1893, fell on Garden City, seventy miles northeast of the Pecos. Although Mayer no longer ranged cattle in Uvalde County, his other Southwest Texas grazing lands also benefitted from early-June showers. The Pecos country, not surprisingly, persisted dry even longer, but even the JM cattle were sloshing through mud within another two months.

No sooner had Circle Dot cattle begun to fatten than they faced another peril—mountain lions. The cats always had hunted the Big Bend, occasionally bringing down a calf or a sickly cow, but the number of killings had now grown to epidemic proportions. In July *The Alpine Avalanche* called for a bounty on the lions, which, the newspaper claimed, played a larger role in short calf crops than most stockmen realized.

Indian Territory also had become dangerous country. Murder and robbery had become commonplace, and authorities seemed powerless to cope with the bands of outlaws responsible. Whether the rise in crime had any bearing on Mayer's decision to seek range elsewhere is not known, but by midsummer he evidently had secured pasturage in the Texas Panhandle. Ample forage throughout Texas now made it feasible to drive again, and in late July John E. Shoemake of the Circle Dot announced that he intended to start a herd of steers and cows for the Panhandle in early August.

Mayer also bolstered his Atascosa County holdings, acquiring an additional section valued at $1,600. He presumably supplemented his deeded land, now totaling approximately 2,000 acres, with leased pastures, for he bought 500 yearling steers from Campbell & Son of Atascosa in late summer and was ranging 700 cattle in the county by 1895. It was the only year he ever declared livestock in Atascosa, although in succeeding years he continued to add to his existing acreage.

On Saturday, November 3, 1894, Mayer arrived in Midland with 1,100 animals bound for the nearby Quien Sabe Ranch, where he had arranged pasturage. Quien Sabe manager Major Ed Fenlon and foreman Barnes Tullous also had been in town, no doubt receiving the cattle. By Mayer's standards, the events of that Saturday probably didn't seem particularly noteworthy; after all, ever since his days in Liberty, he had handled cattle across the breadth of Texas and throughout a host of other states and territories. Yet, within a short while, the Quien Sabe would be virtually synonymous not only with his surname, but with everything that a classic giant-of-a-ranch should be.

Chapter 12

PEÑA COLORADO RANCH.
MARATHON P.O PRESIDIO CO.
TEXAS

M. HALFF & BRO.
STOCK GROWERS AND DEALERS IN CATTLE

PECOS RANCH.
FORT STOCKTON P.O. PECOS CO.
TEXAS

The Quien Sabe

By the time Mayer entered the picture, the Quien Sabe already had the kind of history that was the stuff of legend.

It originated as the "Two Moon" outfit, so-called because its brand—two half-circles, one under the other with open sides down—resembled a pair of thin crescent moons in the sky. By 1881 the Two Moon was ranging cattle along the Rio Grande from below Fort Hancock up to near Alamogordo, New Mexico. Rufe Moore and Barnes Tullous, two respected cowmen destined for prominence in Mayer's cattle business, had connections with the operation. Tom "Black Jack" Ketchum, who later turned to robbing trains, also rode for the outfit.

Eventually, through careless branding or conscious effort, the Two Moon began burning the bottom crescent in an upright position so that the two half-circles almost interlocked. As folklore has it, a stranger, struck by the refined brand's uniqueness, asked a Mexican cowboy what it represented. *"Quien sabe?"* he replied, Spanish for "Who knows?" Soon the Two Moon outfit with the singular brand was known as the Quien Sabe.

In the early 1880s, Quien Sabe manager Edgar B. Bronson, scouting for new range on the Pecos, found twenty miles of east-side river front in Crane and Ward counties immediately upstream of the TX Ranch and Horsehead Crossing. By 1883, Bronson Cattle Company, which owned the Quien Sabe, set up

97

headquarters ten miles above the TX's north boundary and grazed thousands of cattle on the Pecos while continuing to maintain its herd on the Rio Grande.

Like all east-side Pecos outfits, the Quien Sabe suffered the travails of the Big Drift and the drought of 1885–87, double catastrophes that decimated its cattle and grasses. The writing was on the wall by July 13, 1889, when the *Texas Live Stock Journal* made a blunt assessment of the Quien Sabe range: "The broad valley . . . that was a few years ago covered with a thick coat of grass, now only produces mesquite brush and jack rabbits. The grass is all gone." With the Rio Grande equally impacted, the Quien Sabe had no choice but to find new pasturage.

By late July, Bronson Cattle Company had purchased the George Gray range, an expanse of deeded and leased acreage on the Staked Plains centered eighteen to twenty-five miles south of Midland.

"The land was perfectly flat, and one could see for miles in any direction," recalled Ivan Murchison, one-time JM foreman.

Cowhands immediately began sinking wells, for natural water sources were limited. Soon Quien Sabe cattle were on the trail from the Pecos, ready to join their Rio Grande counterparts on the flats above the Midland-Upton county line. By early December, 8,000 to 10,000 head grazed a range described by stockman George B. Loving as the equal of any he had ever seen.

By November 1894, when Bronson Cattle Company and M. Halff & Brother implemented an agreement to graze Halff cattle on Quien Sabe land, the two outfits already stood on common ground. Like the Halff ranches, the Quien Sabe had expanded its country by lease, ranged beeves in Indian Territory, grappled with droughts and low cattle prices, and gained considerable notoriety. Before two more years had passed, the parallels would be even greater—each could rightfully claim, for respective portions of the 1890s, the same brand, herd, and grazing lands.

In the interim, however, Mayer pursued other matters. Sometime in 1895 he presumably sailed to Europe with Rachel and one of their daughters, probably Lillie. He had written an old friend in Lauterbourg of his plans to visit Alsace and had received a reply dated August 8. "You, your wife, and your daughter will be welcome in my house, and how happy I would be to see you again," wrote L. Fromenthal.

Upon his return, Mayer set up a feed yard on his Bee County Ranch to fatten his steers by a technique other than grazing. Never afraid to incorporate new methods, he erected a silo and fed the beeves meal along with the silage—a "great combination," he told the *Texas Stock and Farm Journal* in December.

On February 22, 1896, Mayer significantly enlarged the McMullen County portion of the Prince Pasture. He purchased 10,517 acres, or almost 16½ sections, from W. A. Lowe for $11,874 and other consideration. In McMullen alone, a Prince steer could now range across a staggering 43½ square miles of territory.

The spring of 1896 also saw Mayer return his operation to Indian Territory, evidently for the first time since 1894. In late March he announced his intention to ship one hundred car loads of cattle north of the Red River. These evidently were JM animals, for on April 11 Mayer rendezvoused in San Angelo with his foreman George Noble, who was up from the JM with 2,500 cattle scheduled for transport to the Nations.

Mayer also pastured a significant herd of cattle in Kansas that year, a venture made lucrative by good prices at the fall market. Upon returning from the Kansas City stockyards in early October, he reported that he had sold a large number of Kansas beeves at "very satisfactory prices," according to the *Texas Stock and Farm Journal*.

All the while, he had reflected on the Quien Sabe, for he always "had his eyes open for good land," recalled his son Henry. Furthermore, Quien Sabe foreman Barnes Tullous may have piqued Mayer's interest in the ranch. According to Tullous family tradition, Tullous convinced Mayer and Solomon to buy the Quien Sabe and allow him to manage it on a shares basis; the partnership eventually led to a similar arrangement on the JM, with Tullous' older brother Riley the participant.

By late October Mayer was ready to make an offer to buy. He and Ed Fenlon, manager of Bronson Cattle Company, were in Fort Worth October 27 apparently negotiating a deal not only for the pastures, but also for the Quien Sabe brand and entire herd. On November 4 Mayer closed the sale with the owners of record—Cooper, Hewett & Company of New York—and walked away with 11,000 finely bred cattle and "one of the largest ranches in the West," said the *San Antonio Express*. Only two 160-acre tracts were deeded land, but Mayer assumed leases on almost one hundred sections. The sprawling expanse was completely fenced, a rarity in West Texas at the time, and offered thirsty cattle relief at twenty-two windmills.

Moving quickly, Mayer departed San Angelo November 8 for Midland and the Quien Sabe, where he was to receive the Bronson Cattle Company herd. On November 20, with the tally perhaps complete, the Bronson firm officially transferred the Quien Sabe brand to M. Halff & Brother.

Evidently, the same transaction that brought Mayer the Quien Sabe also yielded him a lease on the 144-section Mallet outfit southwest of Lubbock. Forming a square twelve miles long, the Mallet, which was likewise fenced, extended from near Plains to just inside the New Mexico line.

Like the Quien Sabe, the Mallet already had an involved history. George W. Littlefield, R. A. Houston, and a man named Dilworth had used the Mallet as a road brand for northbound herds in 1877. Six years later, two Connecticut men, Roswell A. Neal and Dwight P. Atwood, joined with Texans Thomas M. Peck and George McWilliams to form the Mallet Cattle Company. The firm's beeves, many of which carried a mallet-like brand, ranged along Morgan Creek in eastern Mitchell County and western Howard County.

Twenty-one months later, Neal, Atwood, and seven other persons incorporated with a capital of $200,000. Atwood set up headquarters near Iatan on the T&P west of Colorado City, but overgrazing and drought laid waste to the range by 1886. Venturing 120 miles northwest, the Mallet secured a vast ranch in Gaines and Yoakum counties in Texas and Lea County in New Mexico.

Although the outfit owned more than six thousand cattle, drought woes and a decline in market prices plunged the firm into debt. By 1893 the Mallet declared bankruptcy and began selling its assets. Apparently, Bronson Cattle Company ultimately gained control of the Texas portion of the western Mallet, which, unsurveyed, actually incorporated a band of New Mexico country three and one-half miles wide by twelve miles long.

Now, both the western Mallet and the Quien Sabe were Mayer's, two immense domains in which to breed and fatten prime steers. The boundaries of the Quien Sabe alone, with the additions of the following years, eventually would encompass five hundred to six hundred square miles across Midland and Glasscock counties. From the north fence—initially seven or eight miles below the town of Midland, but later reaching to its outskirts—a Quien Sabe steer could wander thirty miles south toward the upthrusting JM, and twenty-five or so miles east to a point within ten or twelve miles of Garden City. Interestingly, Mayer would never hold deed to more than fourteen and one-half sections—eleven and one-half in Midland County and three in Glasscock, where the majority of his leased territory lay.

Mayer soon established headquarters sixteen miles west of Garden City at a Glasscock County site referred to as "Section Fourteen." Later, the center of operations was along "Fighting Hollow," eight miles southeast of Midland at a location which previously may have been an upper headquarters.

Mayer's cowboys, who knew the ranch as the "Kin Savvy," soon became familiar with water holes such as Consavvy (Quien Sabe) Lake, nine miles southeast of Midland; Stephenson Lake, another five miles to the east-southeast; and Peck's Spring, site of a line camp seven miles south-southeast of present Sprayberry. Cowhands worked enormous pastures bearing such designations as the Benedict, seventeen miles south of Midland; the Pemberton, presumably on Pemberton Draw in extreme western Glasscock County; and the Swamp Angel, in the vicinity of Peck's Spring. It was an empire under the open sky where, as Mayer's son-in-law Fred Goldsmith wrote in 1905, a cowboy could live a "free, healthy, vigorous life."

Along with the Quien Sabe and Mallet, Mayer fell heir to the services of Quien Sabe foreman Barnes Tullous, whose ability and cow-savvy were unsurpassed. Of Cherokee ancestry, he had been with the Quien Sabe from its early days as the Two Moon outfit on the Rio Grande. After the Quien Sabe had taken root on the Pecos, he had ridden the snaking river as ramrod, a role he continued to fill upon the outfit's relocation to the Staked Plains.

"Barnes Tullous [was] a great cowman," recalled cowhand Lee Bell, who settled in the Midland area in 1887.

Quien Sabe cowboy Young Lee, who rode for Tullous, agreed: "Just being a cowman, Barnes Tullous [was] about the best I ever seen. [He could] work cattle easiest, never got in no hurry, but sure got it done—and was a great horseman."

Tullous was such a gifted rider, in fact, that he once saddled and rode a moss-horned steer several miles across rough country during a Pecos roundup.

"He never choused [needlessly provoked] his cattle," remembered Bob Beverly, another Quien Sabe cowhand, "and he taken good care of his cattle and good care of his horses . . . . He never allowed no man to ride a sore-back horse; if a man had a saddle that hurt the horse's back, the man either got him a new saddle or went to town."

Tullous was also superb with a catch-rope. "Pink Paschal and Barnes Tullous," said cowhand Hugh Campbell, "were the best ropers I ever knew. . . . I saw them in a roping contest. Tullous roped 332 calves and Pink 333 without missing a rope."

Moreover, Tullous possessed keen insight into both human and animal nature. "Barnes . . . could just stand around and figure out what the other fellow was thinking," Beverly recalled. "He could see better at night than any man I ever saw, and he knew every night when them cattle [in a herd] laid down, just about what they was going to do that night [i.e. whether they would rest or stampede]."

A man of so many talents, Tullous also displayed enviable supervisory skills. "He said less—and got more work out of an outfit—than any man I ever saw," Beverly noted. "He'd tell you just what he wanted you to do, and that was all he had to say. . . . I am not the only man that rode Quien Sabe horses that will say Tullous was the most capable man of the Old West when it came to handling men and cattle."

Of prime importance to Mayer, Tullous was above all a master cattle breeder. "He was the first man in that country that started breeding up the herd," related Beverly, who gained deep respect for Tullous' sweeping command of the ranching profession. "He was just A-one . . . the best cowman I ever saw in my life."

The Tullous cowboys had plenty to keep them busy, for Mayer's Quien Sabe generally maintained a herd of 10,000 to

12,000 cattle and branded 5,000 calves a year. With Tullous' careful breeding, Mayer could soon boast of the region's finest and largest Hereford ranch. Meanwhile, Mayer set aside the Mallet, which operated independent of the Quien Sabe, for exclusive use as a pasture for 1,500 to 3,000 steers from the Circle Dot, JM, and Quien Sabe.

"Barnes [Tullous] called it the trash pile for the JMs and the Circle Dot," recalled Beverly, who rode fence for the Mallet. "In them times, why, the T&P road was supposed to be the quarantine line for ticks. . . . [The JM and Circle Dot] would always be coming up in the spring with them herds . . . and turn them loose [on the Mallet] and winter them. They'd shed these ticks . . . and then, in the spring, they'd take them either back, or drive them through to Kansas."

Throughout Mayer's life, he had the good fortune of finding reliable lieutenants to manage the affairs of his ranches. Conversely, anytime there was a job to be done, even a foreman as capable as Barnes Tullous never had to look far to find Mayer ready and willing to step in and work.

Chapter 13

A Son on the Range

Mayer made a circuit through his ranches by buckboard a couple of times a year, accompanied by a valet known as "Old William" or "Negro Bill." During Mayer's days as a foot peddler, William supposedly had acted as his porter. In later years William performed the duties of a manservant and advisor; indeed, inasmuch as daily living matters were concerned, Mayer respected William's opinions above all others and relied on his judgment. They were close friends.

At each of the Halff ranches, the foreman kept a pair of little Spanish mules, a practice that allowed Mayer to secure a fresh team at every stop. Once on the range, Mayer was not hesitant to camp under the stars; the good life in the city never spoiled him for the better one away from it.

"He had this . . . camping outfit, and old William done everything," recalled Bob Beverly, who hired on with the Quien Sabe in 1896. "William fixed his meals, William fixed his bed, and put him to bed. He . . . fixed his teeth for him and . . . just looked out after his general welfare. . . . That was all he had to do—look after the old man, take his hat, and take off his shoes, anything he wanted done."

On the range, Mayer's thriftiness sometimes grated on his dollar-a-day cowboys.

"He wouldn't pay very much wages, and didn't want them

104

to eat much beef," Beverly recalled. "He come to the JM outfit, and ol' Matt Shaw was cooking. . . . Ol' Matt trimmed off a hind quarter, cut steaks off of it, and just threw the bones out. [Mayer] picked them up and told ol' Matt, 'This will make soup.' Old Matt cussed him out [behind his back]—he was always getting cussed out by the boys for just such things as that."

But even in his frugality, Mayer displayed respect for his cowhands. As William made their camp and prepared supper one evening on the Quien Sabe, Mayer puttered around near the blacksmith shop, gathering a nail here, a bolt there. Inside, a new windmill man struggled hard to shoe the windmill team. But the Quien Sabe tools were old and useless, and in disgust the new hand flung a worn rasp in the direction of the scavenging stranger, whom he presumed to be only a passerby. With a vocabulary punctuated by choice epithets, the windmill man expressed his opinion that the Quien Sabe owner was an old skinflint—all within earshot of Mayer.

The windmill man was at the bunk house later that evening when Bill Teele walked in with an announcement: "Well, we got company."

"Who's that?" asked the new hand.

"The man that owns the place."

The windmill mechanic's jaw dropped. "I'll bet that's the old feller who was out here picking up nails and taps and things. I throwed away that horseshoeing outfit—it was wore out and I couldn't do nothing with it. I guess I'll get fired."

A couple of days later, however, Mayer went into Midland, purchased new shoeing equipment, and turned it over to Barnes Tullous.

"You got a good man down on that headquarter ranch down there," Mayer told him. "He takes care of everything good. I see his watering [at the windmills] is all in good shape, and he was trying to shoe his horses and he didn't have anything to work with. Mr. Tullous, we must furnish this man with something to work with."

It was not the only time Mayer rewarded competence. In 1900, when Smith and Mussett of Coldwater, Kansas, received more than 4,000 Quien Sabe steers (road branded with a block) at Pemberton Draw, twenty-eight-year-old Bob Beverly escorted the herd on through the ranch to ensure that no strays fell in with the tallied drove. Upon reaching Mustang Draw on the Slaughter Ranch, Beverly checked the herd and found an unmarked Quien Sabe yearling.

"I cut him out and tied him down and put the block on him," related Beverly. "I taken a receipt from this boss man and turned it in, and it went on through [for payment]."

When Mayer inspected the Quien Sabe sometime later, he approached Beverly, who earned the standard thirty dollars a month.

"I see you found one yearling over in Mr. Slaughter's country that didn't have the block on it," said Mayer. "Mr. Tullous is working men that's looking after Halff's interests. I have recommended, since he is putting a good deal of work on you and responsibility, that he should raise your wages five dollars a month."

Mayer and Rachel also may have rewarded "Old William" for his years of faithful service. The valet probably was William Sheppard, described in 1919 as "a negro man employed in the Halff home." Sheppard was pensioned in Rachel's will; he was to receive twenty-five dollars a month for life.

In the wake of his acquisition of the Quien Sabe and Mallet, Mayer experimented with dehorning as a means of fattening his cattle, testing a theory of the day. In December 1896 he dehorned approximately 1,500 head and lost only ten or so, all of which he attributed to improper care. Impressed with the venture's success, he pondered dehorning every steer on an unspecified ranch the following winter, according to the *San Antonio Stockman*. Whether he did so is unclear, but on another occasion he did instruct a foreman named Franklin—possibly Pete Franklin of the Circle Dot—to saw the horns from a string of rough Mexican steers.

In the spring of approximately 1897, a dozen Quien Sabe cowboys, en route to Coldwater, Kansas, with a cattle herd, gained notoriety in temperance circles. Upon nearing Midland, they held the drove and rode into town with a powerful thirst, only to wind up embroiled in controversy.

"They had the Legal Tender Saloon there, and that's where we all watered-out at," recalled Bob Beverly, who bossed the trail outfit. "Had a good hitch rack in front where we could tie our horses and it was the principal watering place of all the range hands."

They quickly learned that their favorite "water-hole" was on the verge of being shut down—at the hands of fervent prohibitionists holding a county-wide temperance election that very moment.

"It was a big fight," recalled Beverly. "Midland then was a very religious town, and quite a school town, and they had a lot of big speakers lecturing there from everywhere, [evangelist] Sam Jones and a lot of noted men. They put up lots of money to carry it."

By election day, the prohibitionists seemed certain to win. They would have, too, by eleven votes, if the Quien Sabe boys hadn't stopped by to quench their thirst. The twelve voted as a block against the measure and the saloons stayed open—by a single vote.

"It was just kind of agreed amongst all of us that we wanted to continue to water there as we went by," recalled Beverly.

The participation of the Quien Sabe drovers caused an uproar. The prohibitionists contested the election and the court grilled the cowhands one by one on their voting qualifications.

"One ol' boy by the name of Childers," recalled Beverly, "he never could remember anything—couldn't remember who he was nor where he lived nor nothing. And everything they'd ask him, he'd say, 'I don't know.' That's all he'd ever say—'I don't know.' They finally excused him."

When the dust finally settled, the election results stood and the Legal Tender Saloon owed the Quien Sabe boys a few rounds of free drinks.

In the summer of 1897, about the time that rains greened the Circle Dot and the Pecos Valley of southern New Mexico, Mayer and his family left for vacation in New York. He always liked to participate in the fall roundups, but a yellow fever outbreak forced a quarantine of San Antonio and prevented him from returning until mid-November.

In the meantime, Solomon or Rufe Moore followed through on a land deal that brought Mayer's cattle back to Uvalde County for the first time since 1892. On September 21

M. Halff & Brother acquired the 2,594-acre Sabinal Ranch from Fannie Simpson and J. H. James. The spread would grow to 8,606 acres in fifteen months and to 15,248 acres (including 1,476 in Medina County) in five years. Although not Quien Sabe in size, the Sabinal did encompass twenty-four square miles of the Hill Country, ample room for the 800 cattle that roamed the ranch by 1901.

Ever since 1895, Mayer had been busy expanding the JM to include thousands of acres of university land in Crockett County. In all likelihood, he already had relocated JM headquarters to university land along Five-Mile Draw just inside Upton County by the fall of 1897. There, several miles north of Pontoon Bridge, his cowhands had dug a well and apparently erected a structure. Now, though, Mayer was ready to push the ranch's north boundary farther into Upton. Through channels (he was still in New York), he purchased 640 acres near present Rankin from Eliza and Jot Elliott on October 19.

It marked Mayer's first deeded addition to the JM since 1884 and launched him into a buying frenzy in Upton. Over the next eight years, he purchased another thirty and one-half sections in the county, giving him control of Five-Mile and China draws and increasing the JM's deeded property to 53,744 acres, or eighty-four square miles. By evidently the late fall of 1899, his JM patent lands comprised nine pastures and 115 miles of fence. His 253 horses, ten mules, and thousands of cattle could water at thirty-one windmills (fifteen with tanks) and at Wild China Pond northwest of present Rankin.

Unfortunately, a water source from which the animals were restricted after 1897 was the very one which had lured Mayer to the region in the first place—the Pecos. By October 23 cowhands had fenced off the river to protect cattle from its harmful alkali, reported the *San Angelo Standard*.

Although Mayer was serious enough about Judaism that he had joined Solomon in founding Temple Beth-El in San Antonio in 1874 (the two were among only three $1,000 contributors), his religion only rarely played a role in his ranching business. Barnes Tullous and Rufe Moore, two men who held Mayer in high esteem, sometimes referred to him good-naturedly as "that old Jew peddler," but anti-Semitism seems not to have been a problem. Mayer's adherence to Jewish law regarding unclean animals did make for a troubling situation once in the fall of 1897, likely during his stay in New York. Moore, who increasingly handled the details of shipping and selling, found cattle prices so low in Kansas City that he traded 2,700 Osage country steers for tons of bacon. Sometime later, cowhand Bob Beverly picked Mayer up in Midland to wagon him out to the Quien Sabe.

"Well, how are you getting along?" asked Beverly.

"I'm bothered," replied Mayer. "This Rufe Moore—that man's gone crazy. He's lost his mind. He traded a train load of my steers, that you sent to the Osage country, for bacon and shipped it back to San Antonio. What in the name of God would a Jew do with a train load of hog meat?"

Moore, unabashed, leased a warehouse in San Antonio and stored the bacon over the winter. By spring, bacon prices had risen three cents a pound, and he sold it on Mayer's behalf for a handsome profit. The next time Mayer was at the Quien Sabe, Beverly approached him.

"Well, how'd you come out with all that hog meat, Mr. Halff?" he asked.

"I'll tell you, that Rufe Moore is the smartest man I ever seen," bragged Mayer. "You know, that bacon made me plenty of money."

Late 1897 was noteworthy for the first known involvement of Mayer's twenty-three-year-old son Henry in the family cattle business. On November 27 Henry shipped two cars of fat beeves to Kansas City from San Antonio, and three months later he sold sixty young calves to the Saunders and Peel outfit for $12.50 a head. Although he had graduated from business college more than four years before, Henry disdained the idea of life behind a desk and aspired to be a cowboy.

"Henry, when quite a lad," recalled Bob Beverly, who was only two years his senior, "would come to the [Quien Sabe] ranch and ask me to let him ride my horse—many times at night to stand my guard and let me sleep. And, of course, I never objected if the herd was broke in and the weather was good."

But Henry's trail to becoming a cowboy was a rocky one. Presumably sometime after the Spanish-American War, during which he saw duty in Florida as an officer with the First Texas Volunteer Infantry, he helped take a drove of Halff cattle north. After a lingering ordeal through unforgiving country, the herd finally reached a river in Indian Territory or Kansas.

"He said they drove the cattle across the river, and the grass was knee-high," related his son Albert Halff, who was born in 1915. "These cattle were hungry, thirsty, and they'd finally gotten to the Promised Land—and they lay down and died. It was the cold water shocked them."

The incident had a profound impact on young Henry.

"He couldn't believe it," recalled his son. "Dad said he lost his religion—he couldn't believe that there would be a God who would do that. He was never religious after that." (In later life, however, Henry reportedly did attend a Baptist church in Mineral Wells.)

Despite the jading experience on the trail, Henry came to represent the high ideals inherent in many of the cowboys he sought to emulate.

"His outstanding characteristic was his honesty," recalled his daughter Ernestine Freeman, who was born in 1908. "He was known to do what he said he would do. And he was good to his family."

"He was a very gentle man, a very loving man," corroborated his son Albert Halff. "And he was very progressive. He liked new things, new breeds of cattle."

Meanwhile, Henry's brother Alex, almost twenty-nine now, had worked his way up at M. Halff & Brother's dry goods operation. Cashier by 1895, he was now in a position to control the firm someday. Through the Texas Secretary of State's office in January 1898, Alex, Mayer, and Solomon incorporated M. Halff & Brother's merchandising branch with capital stock of $500,000. Mayer, despite his concentration on the cattle industry, became president, Solomon vice president, and Alex treasurer.

As spring 1898 arrived, a dry spell now eighteen months in duration was creating problems in Texas. Although a fine rain on May 9 improved prospects for the JM, some outfits suffered losses. A month later, far-ranging downpours—the best in years in Southwest Texas, said the *Texas Stock and Farm Journal*—

muddied the country from the Sabinal and Prince Ranches west to the Circle Dot. By summer's start, the Pecos River was up and the bordering country sported an unusually fertile mesquite bean crop—important forage for grass-deprived stock. Additional rains in August, some as heavy as four inches, soaked the Pecos range from Barstow to Pecos Spring, adjacent to the lower JM.

Despite the showers' promise, as fall approached, some cattlemen worried about winter grass; the Quien Sabe area had received less than six inches of rain since January 1. The mood was especially gloomy around Fort Davis, where water and grass already were scarce and stockmen feared cattle losses in the coming winter and spring. Pecos cattle, meanwhile, endured well on into November, thanks to better range management on the part of many outfits. "Fully 50 per cent of the cattle have been moved from that country within the past few years," the *Texas Stock and Farm Journal* noted on November 9. "Consequently, those left on the range have double the room which has usually been given them."

The JM cattle soon would have more grazing lands of their own. For several years, Mayer had shown increasing interest in the region between Pontoon Bridge and Midland. His JM range had steadily crept northward through Upton County, while the Quien Sabe had given him control of a chunk of Midland County not far above Upton's northern line. However, two events late in 1898 stand as telling examples of his commitment to this southernmost stretch of the South Plains, where, in the days of Indians and no windmills, few travelers on the Butterfield or Goodnight-Loving trails had tarried.

On December 24 Mayer enlarged the giant JM even more with the purchase of twelve sections on China Draw, in Upton County northeast of present Rankin, from W. D. Lombard Liquidation Company. On the very same day, Mayer sold one of his four major West Texas holdings—the Circle Dot in Brewster County, although he retained rights to the brand. For $17,000, Mayer relinquished Peña Colorado and leases on 75,000 acres in the Maravillas Creek area to W. J. McIntyre.

One storied ranch had ended, but the JM and Quien Sabe had stories yet to tell.

Chapter 14

The Schreiner and Halff Outfit

By the late 1890s, Mayer was at the pinnacle of his success and fortune, yet he never forgot his relatively humble roots in Alsace. He not only maintained contact with friends and relatives in the land of his birth, he also remembered the indigent with gifts of money. One such occasion came at the end of 1898, when he forwarded funds for philanthropic purposes to his old friend L. Fromenthal in Lauterbourg.

On January 24, 1899, upon distributing the money according to Mayer's instructions, Fromenthal penned a return letter. After thanking him "on behalf of all our poor for your great generosity" and expressing hope that God would reward him, Fromenthal reflected on Mayer's own assessment of his lot in life.

"I have read in your kind letter how happy and satisfied you are," wrote Fromenthal, "both with your good health and your business and with the happiness you see all around you in your whole beloved family. You are fortunate, my old friend; it is already very nice to be rich, but one is doubly happy when he has as nice a family as yours, well-situated and with children with good futures."

Fromenthal's letter also indicated that Mayer had made plans to visit Alsace soon.

By early 1899 Mayer long had been an advocate of high-quality bulls, a philosophy that Henry now embraced. Heretofore, Mayer had bred his cows primarily to Herefords, but now he was ready to try another breed. Probably in late 1898, he dispatched Henry to Missouri to buy young bulls, and his son returned by early January with two cars of exceptional registered shorthorns. The two lots, one yearlings and the other calves, were bound for the Quien Sabe.

With an overabundance of Hereford bulls on the ranch, Mayer sold ten of his two-year-old Herefords to Henry Packingham of Dryden for $45 a head in March. Two months later, Mayer sold Baldridge Brothers of Indian Territory eighty more Quien Sabe bulls for the outfit's Pecos County ranch.

Fortunately for Mayer and other cattlemen, the expected cattle losses of winter and spring failed to materialize. By late March 1899, green grass in the Pecos bottom land and around the lagunas was already three to six inches high, and soaking rains a month later along the lower JM put Live Oak Creek on a rise. But the Pecos never would be a wetland, and by late August several Grandfalls-area outfits prepared to abandon a river that now offered little more than ripe mesquite beans.

"The bulk of stock . . . are living on grass made prior to 1897," observed a Grandfalls correspondent for the *Texas Stock and Farm Journal*. "It is an old saying that it is a wise dog who walks out when he sees preparations ripe for being kicked into the street."

At the same time, however, the correspondent pointed out that the region's very aridity was, in one respect, an advantage: "Grass does not rot here like it does in a damp country. It simply dries up and turns to hay on the stalk; stock eat and thrive on it all winter and sometimes the rancher will have a lot cut, whatever may be the time of year."

Predators, meanwhile, offered Pecos and Staked Plains cattlemen no benefit whatsoever. Wolves in particular had always been a nuisance, especially along the river. Loafers, as cowboys called them, often ran in packs of five to ten animals and posed a threat even to mature cattle and horses. In 1895 the *Texas Stock and Farm Journal* reported a "regular war. . . being waged against these raiders . . . [which form] a bold, ravenous horde that roams the prairies."

For years outfits had hunted, trapped, and poisoned the

canines, yet they seemed as plentiful as ever. In the summer of 1898, wolf depredations had reached an alarming stage. E. W. Clark, who ranched in the Quien Sabe area, had reported the loss of seven yearlings in one week to wolves, which, according to the rancher, were growing more numerous every year.

On May 8, 1899, Crockett County stockmen finally banded together in an effort to exterminate loafers and other predators. From a tax of a half-cent for every bovine or horse declared by a rancher in the county tax roll—in Mayer's case, $22.25 for 4,450 head—the stockman's group offered bounties ranging from $1 for a bobcat to $15 for a mature lobo.

Sometime in 1899 came evidence that Mayer and Solomon were drifting apart. Over the years, Solomon had stayed active in Alamo National Bank, serving on the board of directors. However, Mayer chose in that year to help organize another San Antonio bank, City National, which was chartered with a capital of $100,000. By August of 1900, Mayer would acquire a large interest in City National and assume the presidency.

The fact that Mayer had established a rival institution may have grated on a relationship possibly already strained. One might speculate that Solomon—who had always been no more than M. Halff's "brother" in the cattle, dry goods, and wholesale worlds—long since had tired of playing subordinate. As early as May 12, 1867, Solomon had taken Mayer to task for criticizing him about a business matter.

"As far as my pretensions to knowledge of goods, they never were very great, and I expected to make some mistakes in my first purchases," Solomon wrote from Lauterbourg. "After having made such unpardonable mistakes yourself, you might only be a little less emphatical in your expressions. . . . With such a letter from you, it is pretty difficult to take your advice in same letter to take things easy and not let business trouble me."

Still, in the absence of documentation, any suggestion of a rift between the brothers in 1899 can only be conjectural.

Almost from the start, Mayer sailed a rough sea in the banking world. By his own words, he plunged in with vim. Confident that City National could quickly muster $500,000 in deposits, he expended his own money toward the cause. Within a matter of weeks, however, "schemes were concocted against me, and . . .

the bank was threatened to come to an end," he recalled in a 1904 letter to J. D. Anderson, the cashier. Anderson, in fact, may have precipitated the threatened takeover with a request for higher wages.

"I considered your demand too hasty," Mayer told Anderson, "and so declared myself, and then and there, plans had been on foot to have me ousted."

Nevertheless, Mayer came to the rescue of the bank, he said, "in order to protect the good name of others as well as my own." Although the details are unclear, the matter apparently did irreparable damage from a public relations standpoint.

"There and then," continued Mayer, "the bank came into disrepute, displacing some of the best citizens of San Antonio from the directory. This gave all kinds of opportunity for talk to our competitors. Since then the bank has lost its prestige, and from that time it has been much to my sorrow to have become connected with it." Then Mayer expounded a philosophy that had carried him from pack peddler to cattle and dry goods magnate. "No institution," he wrote, "can be progressive when it stands still."

Even at age sixty-three in 1899, Mayer continued pushing the envelope, whether in his cattle business or his personal affairs. In May he and Rachel embarked by train for New York, where after a week of evidently visiting their daughter Hennie, they set sail for Europe. He presumably visited L. Fromenthal in Lauterbourg as planned, but an illness otherwise marred his summer abroad. After Mayer and Rachel boarded a steamer for America, his cousin C. Dreyfus wrote him from Paris and expressed hope that he had recovered during the voyage home.

Mayer's West Texas herds, enormous before, now had reached mammoth proportions, as reflected by his summer tally of 4,106 steers, 9,571 JM cows and yearling heifers, and 3,400 calves. At peak numbers that season, his JM herd alone amounted to an astounding 19,020 mixed cattle. (Interestingly, he declared only 8,300 head, or forty-four per cent, to tax officials.) Moreover, he was now shipping an incredible 20,000 cattle to points north every year, according to the *Texas Stock and Farm Journal*.

During the Hanukkah season, Mayer again sent money to

the poor in Lauterbourg by way of L. Fromenthal, who distrib-
uted it in stages to seven parties. "I never give everything to
them at one time," Fromenthal wrote him. "I am keeping part
of it for Purim [the feast of lots, a Jewish holiday], and perhaps
a tiny bit for the Easter holidays."

In the spring of 1900, Mayer had a herd on the trail from
the Mallet, approximately 2,500 two-year-old Quien Sabe steers
headed for the town of Canyon in the Texas Panhandle. There,
Henry would be waiting to oversee delivery to buyers George
Cowden and George Pemberton, who wanted to ship the herd
for Montana by May 28. Mayer received $25 a head, an amount
he deemed acceptable but not worthy of celebration.

Stampede and incessant rain bedeviled the drive, but trail
boss Lum Arnold, aided in particular by Bob Beverly, managed
to bring the herd into Canyon on time. Cowhand Sam Preston,
who happened to be in the city, was awed by the beeves' quality.

"These were the best-looking cattle I had seen up to that
time," recalled Preston, who became Quien Sabe manager in
1907.

The primary reason was Mayer's ongoing commitment to
excellence in breeding. Significantly, after a year in which he
had shied away from Hereford blood in favor of shorthorn, he
returned to familiar ground in March 1900. He evidently sent
Henry to a Kansas City auction, where the twenty-five-year-old
purchased twenty-two choice registered Hereford bulls, each approaching two years of age, for $225 each. By late March Henry
had delivered the 1,200-pound animals to the Quien Sabe.
Henry also journeyed to Iowa sometime during this period and
acquired Hereford bulls for $250 apiece.

Although Mayer reserved the gentler grasslands of the
Quien Sabe for his prize breeding stock, the rugged canyons
and far-reaching drainages of the JM also bore the hoof prints
of progress. By the early 1900s, sixty percent of the JM cattle
were purebred Herefords and the remainder Hereford-longhorn crosses.

With so much invested in his bulls, Mayer never permitted
them to forage for themselves during winter.

"The bulls were cut from the herd, put in a separate pasture, and fed," recalled Ivan Murchison, who was JM foreman

early in the twentieth century. "During this period the bulls were as sweet and nice to each other as Sunday school boys. It was a different story in the spring when they were returned to the herd. Many times at the watering place, the line riders had to use their bull whips to separate two fighting bulls to keep them from killing each other."

Mayer, now sixty-four, was increasingly delegating the supervision of the Quien Sabe, JM, and Mallet to his son, Henry. On May 12, 1900, Henry returned to San Antonio after spending much of the previous four or five months on the ranches, where a favorable winter and spring had nurtured an unusually large calf crop.

Mayer, however, was not the type to sit idly and watch the world rush by. In late summer and fall, he closed several important land deals that greatly expanded his range. On September 4 he added 561 acres to the Sabinal Ranch by purchase, and on November 5 he secured a deed from Catharine Rhomberg for three sections in Upton County. Five weeks later, he extended the Sabinal into Medina County by acquiring 1,476 acres from Baldwin Hufty.

But it was a joint purchase on December 7 with cattle magnate Charles Schreiner of Kerrville that brought the last of Mayer's great ranches into his possession. Known initially as the B. L. Crouch outfit and later as the Schreiner and Halff, this Frio County spread encompassed ninety-one sections on both sides of the Frio River immediately north of its confluence with the Leona River, which formed part of the south boundary. Both river bottoms, and that of Buck Creek, were majestic with live oaks, hackberries, hickories, and pecans. The lowlands of the flood-prone Frio sometimes swelled to a mile wide before giving way to sandy flats on the east and gravel ridges on the west. The spread's rich soil nurtured nutritious grasses, and its 375 cultivated acres had yielded an average of thirty-five to forty-three bushels of corn and one-half to one bale of cotton per acre.

Completely fenced, the Crouch offered eleven pastures, each with its own water source, and its location on the International and Great Northern Railroad provided quick and easy shipping. From Mayer's standpoint, the range probably seemed a paradise compared to the harsh canyons and inconvenient plains of West Texas.

M. Halff & Brother and Schreiner presumably paid British and American Mortgage Company the asking price of $120,000, including four notes totalling $38,292 and a final note, due December 7, 1904, for $4,660. They also acquired a supplemental 210-acre tract from Charles H. Mayfield.

The three principals—Mayer, Solomon, and Schreiner— each had undivided interests in the Crouch, with Schreiner owning fifty percent and Mayer and Solomon splitting the remainder. Soon after the purchase, however, the brothers took the first step toward dissolving their long-standing business relationship. On January 30, 1901, Mayer acquired Solomon's share of the Crouch by deed, making him equal owner with Schreiner.

Mayer had much in common with his new partner. Schreiner, two years Mayer's junior, also had been born in France. He had moved to San Antonio in 1852 and had entered the livestock trade in Kerr County in 1857. Twelve years later he had established a merchandising business that eventually had given him the means to found the Charles Schreiner Company, a leader in the ranching and banking industries. In 1880 Schreiner had purchased the famous YO Ranch in the Hill Country, and by 1900 he owned more than 600,000 contiguous acres between Kerrville and Menard.

With their new partnership, Mayer and Schreiner immediately assumed a position of power and influence in the Texas cattle industry. With two such extraordinarily capable cowmen at the helm, the Crouch could do nothing but prosper. By 1902 it grazed 2,200 cattle and 45 horses on the Frio, and following years brought even greater livestock numbers—4,250 cattle, 48 horses, and 800 sheep by 1905. The ranch's boundaries also expanded during that span, stretched by the purchases of ten additional sections. In 1905 the Frio County tax office put it all in financial terms, appraising their operation at $236,695, up sixty-one percent from 1902.

In 1917 Schreiner deeded his interest to his daughter, Lena Partee, who in turn sold it to the Halff family in 1924. Of all Mayer's ranches, the Schreiner and Halff alone remained in his descendants' control in 2000.

The dawning of the twentieth century in 1901 finally brought the cattle industry abreast of Mayer, a cowman who never had let himself be rooted to nineteenth-century ways. But even while he continued to pursue ranching excellence, by year's end he would face a troubling development, both professionally and privately.

Chapter 15

Brothers Part Ways

Before the travails of late 1901, Mayer found time for pleasure. In the summer, he and Rachel evidently returned to Europe, where they planned to spend time on the Rhine and visit L. Fromenthal in Lauterbourg.

For the family of one Quien Sabe cowboy, however, summer was far from pleasant. The year before, Oscar Midkiff had hired on and moved his wife Lillie and their three young children to a dugout at Peck's Spring. As Midkiff rode away on cow work, foreman Barnes Tullous promised to dispatch food for his stranded family.

For whatever reason, however, the supplies never reached Peck's Spring, forcing Lillie to desperate measures. On the Fourth of July, Midkiff rode up to find her standing at a pond, in her hands a crude fishing pole with a bent pin for a hook.

"Lillie, what on earth are you doing?" he asked.

She broke into tears. "I'm trying to catch a fish—my children are *starving*. I've got half a cup of weevily corn meal and a cup of dried apples."

Midkiff immediately rode back to headquarters, demanded an accounting from Tullous, and quit in anger.

By then Tullous' brother Riley, who was born in 1857, had taken the reins of the JM. On business in San Angelo in late July, Riley reported that steady rains had rendered the JM "as green as a wheat field." He also informed the *San Angelo Standard* that the JM had turned down an offer of $18 a head for its 3,000 yearlings.

By the fall of 1901, Mayer and Solomon had been inseparable in many ways for much of the past forty-four years. As young immigrants reunited in a strange land, they had reaffirmed their strong family ties and had benefitted from each other's expertise. Even as they had risen to the apex of success, their bond had remained so firm that they had even built their respective homes a stone's throw apart. But now they chose to go their separate ways, with Mayer acquiring Solomon's half-interest in all their lands and herds, with only one exception.

In a short document signed on December 7, the brothers dissolved their ranching partnership, with "S. Halff retiring from said business." An appended document, however, noted that the Prince Pasture and herd would remain undivided with Mayer as manager. In a third paper, handwritten in apparent haste on the same day, Solomon resigned as vice president and director of M. Halff & Brother Incorporated, the merchandising branch of their operation.

The split was final and virtually clean. One story holds that the brothers remained civil to one another during the dissolution until they came to the last item to be divided—a new buckboard with team. Unable to agree on who would get it, they parted on bitter terms and never spoke again.

Solomon's subsequent actions lend credence to this report of an unfriendly breakup. Rather than going quietly into the night, he immersed himself in direct rivalry not only against Mayer's banking interests, but against the very dry goods firm that both of them had nurtured for three decades.

Initially, Solomon turned his attention to Alamo National Bank, where he had become vice president by 1902. Two years later, he became controlling stockholder and president of A. B. Frank Company, one of M. Halff & Brother's chief competitors. On January 3, 1905, he sold his share of the Prince Pasture to Charles Schreiner for $27,346. By January 10, when Alamo National elected Solomon chairman of the board and first vice president, it was obvious to everyone that his relationship with Mayer was strongly adversarial.

Whatever Mayer's feelings toward his brother in later years, he seemed to take the breakup in stride in 1901. In addressing the needs of Lauterbourg's poor again late that fall, Mayer sent L. Fromenthal a check for one hundred francs and wrote of his happiness and good health.

One reason for his high spirits may have been the fact that Solomon's departure cleared the way for Mayer's older son to move up in M. Halff & Brother. Mayer had already granted Alex power of attorney; now he placed him in active control of the company, keeping himself free to pursue other interests. Under Alex's tutelage, the company flourished to the point that stockholders were able to sell it to A. B. Frank Company for one million dollars in 1929, only weeks before the stock market crash.

Henry, meanwhile, soon registered the Quien Sabe brand in Midland County under his own name, which suggests that he had all but assumed total supervision of the ranch. In a letter probably written on December 28, 1901, he also provided estimates of livestock on the JM, Quien Sabe, and Mallet for Mayer's use as "a guide should you decide to make an offer on the property." Evidently, he or Mayer entertained a passing notion to sell. Henry assessed the Mallet cattle herd at 2,000 head, the Quien Sabe at 10,750, and the JM at 12,300. A 300-horse remuda boosted his total valuation of livestock to a lofty $422,000.

Oddly, Henry's assessment of land values seems extraordinarily low—$5,000 for the Mallet, $20,000 for the Quien Sabe, and $40,000 for the JM. Possibly, he referred only to existing leases, not to deeded land; nineteen years before, Mayer and Solomon had paid a full $80,000 for the JM's river frontage alone.

But Mayer apparently had no intention to liquidate the ranches. Indeed, a report that he handled 40,000 steers in 1902 is evidence that, even without Solomon's financial backing, Mayer's ranching strategy had grown only more grandiose. That year, his plans included Indian Territory, to which he freighted eighty cars of cattle after loading them at San Angelo starting March 31. That same month, the JM sent 2,100 raw-boned longhorns north from the lower Pecos for Portales, New Mexico, where rail rates to Kansas and other points north were cheaper.

The five- to six-year-old cattle, which weighed an average of seven hundred pounds apiece, proved troublesome on the seventeen-day drive. A passing train stampeded the herd in Odessa, and the terrified animals wreaked havoc with the town. For a couple of weeks thereafter, the drive went smoothly, but more disaster awaited on the New Mexico plains. At a point three days shy of Portales, lightning struck cowhand Bob Williams as he rode out on night guard. Knocked unconscious, he fell from his addled horse, and the boogered animal bolted and stampeded the herd a second time. After a half-day roundup, the nine-man crew pushed the cattle on to Portales, the destination of several JM drives in the early 1900s.

Mayer ushered in the close of 1902 with yet another contribution to Lauterbourg's needy via L. Fromenthal. He also informed his old friend that he and Rachel were planning a trip abroad for the summer.

In the interim, however, there was plenty of ranching business to conduct. About this time, Mayer financed Asa Jones, a former Quien Sabe cowhand, with $10,000 for a horse-trading venture. Over the next two years, Jones acquired horses in the region and resold them at a handsome profit in San Antonio, a city with a reputation as the biggest horse market in the world.

At some point Mayer had invested heavily in Drumm Commission Company, a livestock brokerage firm with offices in Kansas City, Chicago, and St. Louis. In the spring of 1903, however, he decided to cast his lot with Southwestern Live Stock Commission Company and joined its board of directors. As a result, he approached Drumm Commission and tried to liquidate his considerable stock in the firm. Presumably he did so, even though company president A. Drumm could offer him no more than an admittedly paltry $22,500 for his 250 shares on April 13.

Meanwhile, Mayer was growing testy with Henry over the way his younger son was handling his West Texas ranch affairs. After all, Mayer had been dealing in cattle since his days in Liberty, and Henry's actions were suddenly flying in the face of all those decades of tried and proven methods. During a New York stay, Mayer berated his son by letter, only to receive a stinging reply.

"For two months you have been quarreling with me by

mail," Henry wrote, "first because I did not buy any steers and afterwards because I did." This was almost certainly a herd of 1,395 yearlings he had purchased through Max Mayer of Sonora, Texas, for $16,740 in midsummer. "I suppose your reason for this has been to get up an argument (just because you'd nothing else to do) but I have been too busy to argue. How do you expect your ranches to pay lease, or expense, or interest without any cattle? You know better than that! You just want to tease me and worry me because you think I have nothing to do. I have plenty worry and aggravation without keeping up a continual argument with you."

Another point of contention was a flock of 1,500 goats Henry had purchased from Val Verde Land and Cattle Company for one dollar a head on July 18.

"M. Halff & Bro. had no goats," Mayer reminded him by letter. "They had a flock of sheep once and gave them away."

By now, twenty-nine-year-old Henry had enough self-assurance to offer his sage father a blunt assessment: "Had we kept the ewes out of that flock [of sheep], we would now be making all the ranch expenses and lease out of sheep. I did not know that we were expected to do just as M. Halff & Bro. did twenty years ago. I have never said nor insinuated nor thought that you were old fashioned or in your dotage, as you put it. I simply thought you wanted arguments."

Despite their differences, Mayer gave Henry 1,500 cattle that year, along with four deeded sections of Quien Sabe land in Glasscock County. Glasscock tax officials valued the acreage at $3,840 and the herd at $12,000.

Back in the early days of ranching on the Pecos, ticks had never posed a direct problem for the hardy longhorns that ruled the range. As cattlemen bred more and more high-grade blood into their herds, however, they found their cattle increasingly suffering the ill effects of the parasites, aside from their role in Texas fever. On the JM, Mayer's solution was to build a one thousand-dollar dipping vat in the summer of 1903. Upon its completion, JM manager Riley Tullous planned to fill it with five feet of water and one foot of Beaumont oil and dip 10,000 steers. As usual, Mayer was a step ahead of his contemporaries.

"The proceedings," said the *Texas Stock Journal* on

August 26, "are being watched with interest by stockmen of that section, who will 'go and do likewise' should the experiment prove a success."

Mayer may have had methods of intervening in regard to ticks, but the skies were another matter. Along the Pecos and throughout much of the surrounding area, a fiery sun torched the land day after day beginning in early July. By the time Henry sold 105 Quien Sabe calves in Kansas City in early November, grass was scarce near Big Spring and north of Midland, creating concerns for the winter. By Thanksgiving, the Pecos range also was in poor shape, and the onslaught of winter a month later only exacerbated matters.

Although livestock in some areas of West Texas wintered well, by late February 1904 skeleton-like cattle walked the Pecos under skies still cruelly barren. The Quien Sabe received a three-inch rain in early April, but elsewhere the drought lingered unimpeded into May; Crockett County and the lower JM had now been without precipitation for ten hard months. "Cattle . . . have got to have grass, and have it quick," observed the Crockett County correspondent for the *Texas Stockman-Journal* on May 11. "What grass there is on the range was matured nine months ago and is almost worthless."

Cattle were so thin and weak that, even if rains came immediately, losses were all but inevitable. "There is going to be a whole lot of tailing up, pulling out of bog holes and skinning to do," noted the *Journal* writer.

With a big die-up imminent, a thunderhead rose up out of the northwest on May 17 and deluged Crockett County, the region's first significant precipitation in ten months. Runoff filled the water holes and rushed down the arroyos with such force that the Tom Patrick ranch lost a fence. Nevertheless, said one area resident, the patter of rain "was sweet music to our people."

During the next week, scattered showers fell across the Pecos Valley and west to the Davis and Guadalupe mountains. The drought, now bent but not broken, finally yielded along the Pecos on May 25, when a general downpour soaked its valley and flats. It put the river on a "booming" rise, reported the June 1 *Texas Stockman-Journal*, and caused several washouts along the Pecos Valley Railroad north of the city of Pecos.

In early June, the Mallet and Quien Sabe were blessed with

such an encouraging rain that the *Midland Reporter* was bold enough to make a pronouncement: "Now we can say with all consistency that the drought has been broken." Follow-up precipitation in July did nothing to dispute the newspaper's contention. Jim Crenshaw, ranching in south Midland County near the Quien Sabe and upper JM, reported a downpour unlike any he had ever seen.

"Prairie dogs, cotton tails . . . jack rabbits . . . and two coyotes were found drowned," said the *Midland Reporter*, adding that Crenshaw even had to "put life preservers on the calves to keep them safe." Perhaps similarly tongue-in-cheek, Crenshaw also disclosed that his cattle had developed quite a taste for the thousand-legs, or centipedes, that the rains had spawned.

For a country baked dry, the benefits of this season of showers were profound. By mid-July, West Texas had a "beautiful spread of green" and cattle were sleek and putting on flesh, reported a Midland newspaper.

Once again, Mayer had endured the uncertainties of the Pecos, and this time he also had introduced Henry to its unfathomable ways. Perhaps the son had come to understand as the father did that there were three things this uncompromising land required of a cowman: confidence, stamina, and, above all, unflagging patience.

"Things always come out on top in West Texas if we just wait long enough," reflected the *Ozona Texan* that summer. "After all, this is the best country on earth."

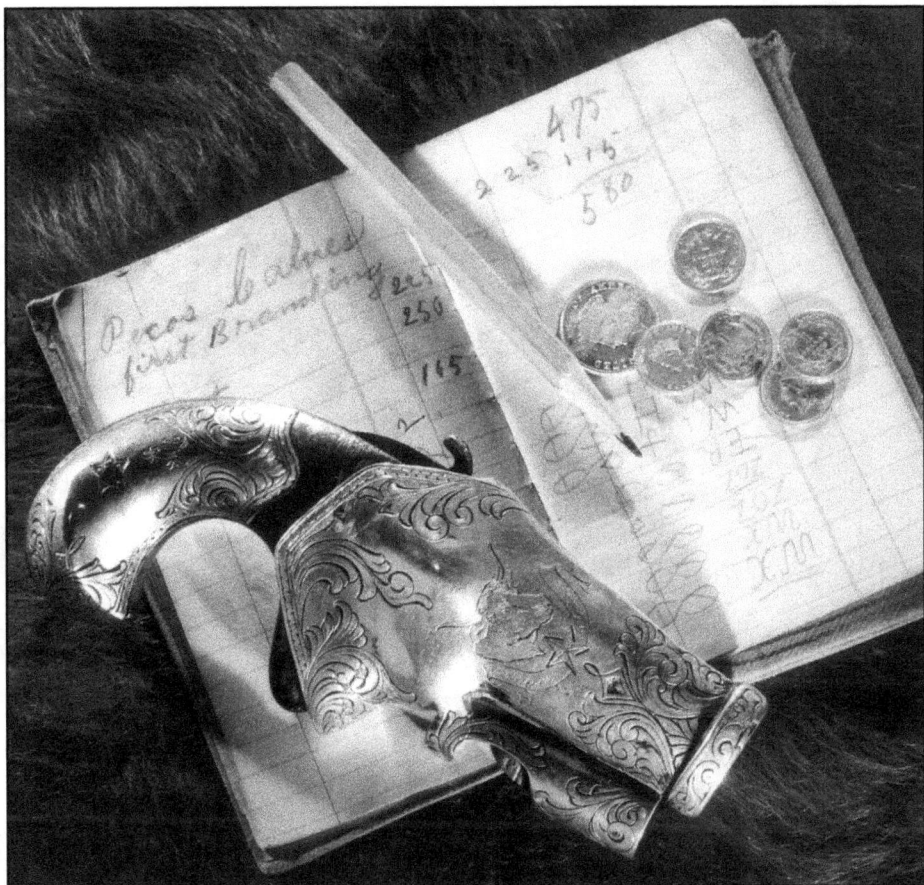

Mayer Halff's derringer, dating to circa 1872 or 1873.
— Courtesy Donald Yena

Alex Halff, Mayer's son, at age twelve.
— Courtesy Alex H. Halff

Mayer Halff's son Alex as a child.
— Courtesy Alex H. Halff

German - English School.

Testimonial of *Alex Halff*

For the term from *Sept 4th* to *Dec. 23rd* 1876.

Behavior, *very good*

Attendance, *regular*

Industry and progress, *very good*

J. E. Cotton

R. Vielant

Alex Halff's German-English School report card for 1876
— Courtesy Alex H. Halff

Henry M. Halff as a child.

— Courtesy Alex H. Halff

Henry M. Halff, age thirty, with son Mayer in San Antonio.
— Courtesy Alex H. Halff

Mayer Halff, Rachel Halff, and their son Alex in 1898.
— Courtesy Alex H. Halff

The Dripping Springs region in the Organ Mountains.

— Courtesy Alex Halff

W. W. Cox (far right) and others on the Cox Ranch.
— Courtesy University of Texas at El Paso Library, Special Collections
Department, W. W. Cox Family Papers

*The Cox Ranch near San Augustine Pass in the Organ Mountains
of New Mexico.*
— Courtesy University of Texas at El Paso Library, Special Collections
Department, W. W. Cox Family Papers

W. W. Cox (standing, center) branding a calf. Nearby are (left to right)
Mr. Coe, Sterling Rhode, Jim Hester, and Ed Cox.
— Courtesy University of Texas at El Paso Library, Special Collections
Department, W. W. Cox Family Papers

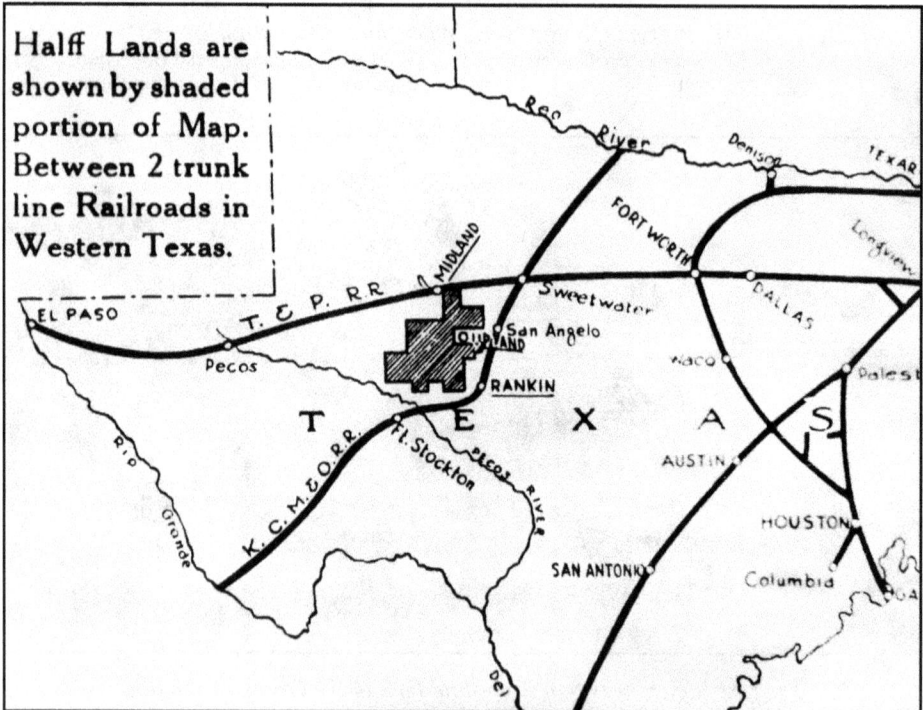

Advertisement showing extent of Quien Sabe and upper JM.
— From The Halff Lands: 80,000 Acres in West Texas

Quien Sabe cowboys in Midland about 1906: Ben Driver, Fred Truelove,
Rabe Preston, Gill Haynes, James Curry, Bob Preston, Cleo Gaither.
— Courtesy Midland County Historical Museum

Quien Sabe headquarters in Glasscock County in 1909.
— Courtesy Nancy McKinley

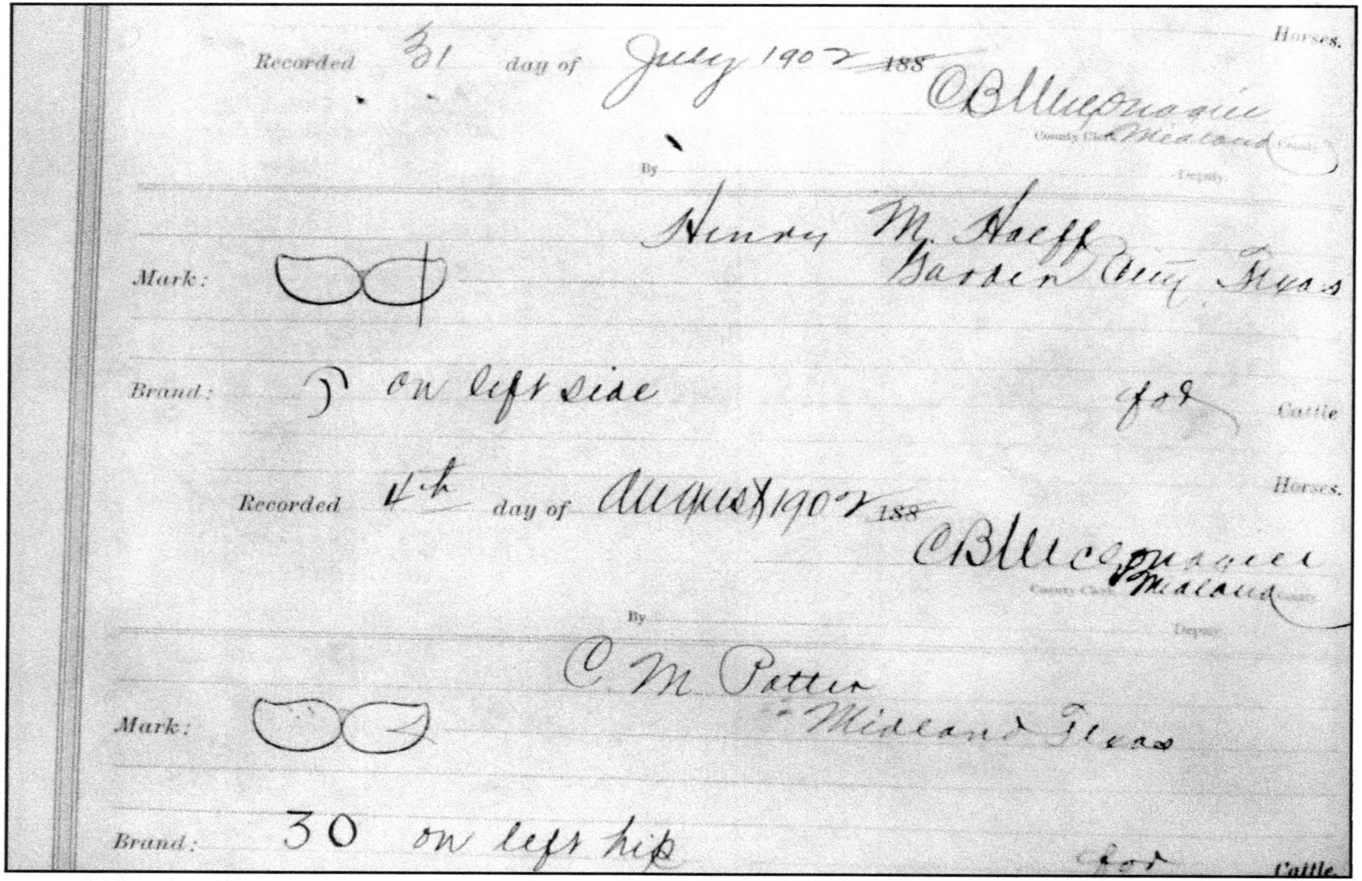

Horses.

Recorded ___ 31 ___ day of ___ July 1907 ___ 188 ___ O. B. McClintic
County Clerk, Midland County

By _____ Deputy

Henry M. Halff
Barston [?] Wing, Texas

Mark:

Brand: ___ D on left side

Cattle

Horses.

Recorded ___ 4th ___ day of ___ August 1907 ___ 188 ___ O. B. McClintic
County Clerk, Midland County

By _____ Deputy

C. M. Cutter
Midland, Texas

Mark:

Brand: ___ 30 on left hip

Cattle

The Quien Sabe brand as recorded by Henry Halff in Midland County on August 4, 1902.

Barnes Tullous
— Courtesy Nita Stewart Haley
Memorial Library,
Midland, Texas

Bob Beverly with his horse Cleo in 1915.
— Courtesy Nita Stewart Haley Memorial Library,
Midland, Texas

Pay to Order of   *No.*

SERVICES

Quien Sabe Ranch.

from

to

$

John Routledge & Son, Printers, San Antonio.

FORM 1.

190

Quien Sabe Ranch,

To   Dr.

For services from *to*

days at $ *per*

as

Amount advanced previously, - - - -

- *Balance,* - - -

I hereby certify that above amount is correct, and services were rendered as stated

No.

M. HALFF & BRO.,
San Antonio, Texas.

Pay to   *ot Order,*

Dollars,

for services rendered as stated above.

$

This Draft not good if Detached.

Quien Sabe Ranch check and pay stub.

— Courtesy Richard Cauble

The Preston brothers, including six one-time Quien Sabe cowhands, in 1921.
(Top) Sam, Rabe, Tom, Barney; (bottom) Kirb, Bob, Ed.
Only Tom never rode for the outfit.
— Courtesy Nancy McKinley

Bob Preston on a Quien Sabe animal in 1909.
— Courtesy Nancy McKinley

Quien Sabe cowhands in 1906: (top) Pat Talmadge, Charlie Brown, Bob Preston; (bottom) Joe Stokes, Bud Mulligan.
— Courtesy Nancy McKinley

Snake Haines, Rat Brown, Dick Moody, and Bob Preston on the Quien Sabe in 1909.
— Courtesy Nancy McKinley

Heading and heeling on the Quien Sabe.
— Courtesy Nancy McKinley

Rabe Preston (left) earing-down a Quien Sabe bronc for Snake Haines in 1909.
— Courtesy Nancy McKinley

Henry Halff
— From The Halff Lands: 80,000 Acres in West Texas

Silos and Hereford cattle on the Quien Sabe.
— From The Halff Lands: 80,000 Acres in West Texas

*"Hero" and "Dove," part of Henry Halff's registered show herd
from the Quien Sabe.*
— From The Halff Lands: 80,000 Acres in West Texas

Henry Halff and Quien Sabe water well near Midland.
— From The Halff Lands: 80,000 Acres in West Texas

Herefords on the Quien Sabe Ranch in February 1900.
— Courtesy Alex H. Halff

Henry Halff, standing, and Virgil Judkins, horseback,
at a Quien Sabe or JM cow camp.
— Courtesy Midland County Historical Museum

Registered
Hereford Bull
Beau Homage
No. 510,611
Age 36 Months
Weight 2,200 lbs.
Bred, Fed and
Shown
by
HENRY M. HALFF
Midland, Texas

Henry Halff's business card.
— Courtesy Nancy McKinley

Roundup on the Quien Sabe or JM.
— Courtesy Nancy McKinley

Roundup on the Quien Sabe or JM.
— Courtesy Nancy McKinley

Henry Halff's oil pull tractor, used for road grading.
— Courtesy Midland County Historical Museum

Charles Schreiner
— Courtesy Nita Stewart Haley
Memorial Library,
Midland, Texas

*Schreiner and Halff Ranch manager John Little (right)
with Sam Stern in 1922.*
— Courtesy Alex H. Halff

*John Little (left), Alex H. Halff, Arthur Witchell, Sam Stern, Mrs. Lester
Pranglin, and Lester Pranglin at the old Schreiner and Halff Ranch in 1921.*
— Courtesy Alex H. Halff

Mayer Halff's cemetery plot.
— Courtesy Alex H. Halff

Grave of Rosa, Mayer's infant daughter.
— Courtesy Alex H. Halff

MAYER HALFF
FEBRUARY 7, 1836
DECEMBER 23, 1905

*Mayer Halff's grave
marker.*
— Courtesy Alex H. Halff

Halff family portrait: (back) Lillian Goldsmith, Frederick Goldsmith, Henrietta (Hennie) Halff Goldsmith, Mayer Halff, Rachel Halff, Gertrude Goldsmith; (front), Ruth Goldsmith, Henrietta (Hennie) Halff Goldsmith.

Chapter 16

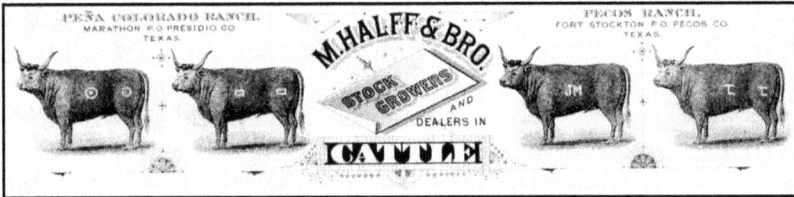

End of the Trail

By the time Mayer approached his sixty-eighth birthday on February 7, 1904, he already had lived several lives. He had been a Jewish boy tending his father's cattle in the lush Alsatian valleys, then a young immigrant peddling his way across the Texas Gulf Coast. In early manhood, he had played the role of merchant and built a business that had swelled to gargantuan proportions. Then the hides and horns of Texas cattle had summoned him to his ultimate destiny as czar of a ranching domain stretching across several states or territories. He had veritably lived the American dream, building on a foundation of integrity and determination to achieve something greater than wealth —respect.

"No one," remembered William H. Porter, president of New York's Chemical National Bank, "stood higher in our esteem."

In Mayer, both the affluent and the poor found a man truly worthy of honor, and he did nothing but strengthen his legacy as he neared the end of life's brief trail. His kindness and generosity burned brightly yet again on December 12, 1903, when he contributed once more to the poor in Lauterbourg in a letter to L. Fromenthal. He also informed Fromenthal that he and Rachel would visit France again in the summer.

Mayer's admirable traits had even earned him a title of

respect in the cattle industry. The *Texas Stockman-Journal* of April 12, 1904, reporting that he had received word of a three-inch rain on the Quien Sabe, referred to him as "Col. Meyer [sic] Halff."

With increasing age, however, came health problems. Although the exact nature of his ailment is not clear, it induced him (and likely Rachel) to travel to Atlantic City, New Jersey, for the summer. "This is an ideal spot and my health has improved very much since I came here," he wrote on July 5. "We expect to remain here until the 24th." Whether he and Rachel proceeded to Europe as planned is not known.

Frustration, perhaps exacerbated by poor health, led Mayer to step down as president of City National Bank that summer and offer his stock for sale. For once he had failed to make a success of a business endeavor, while Solomon soon would be elected chairman of the board and first vice president at rival Alamo National Bank.

"It is not for past grievances that I wish to retire," Mayer wrote cashier J. D. Anderson. "I wish to withdraw from the institution simply for not having been able to have made more of a success than this bank has made, a bank which has been, in a measure, the laughing stock of our competitors . . . through the actions of its officers."

Out in West Texas, meanwhile, Henry had assumed firm control of the Quien Sabe and JM. About this time, he relocated the JM headquarters from Five-Mile Draw to a two-room rock house on China Draw, eight miles north-northwest of present Rankin. Built in 1880 by Dr. George Elliott of San Antonio, the sixteen-by-thirty-foot structure had three entrances, a back port-hole, and two rooms separated by a fireplace. A couple of hundred yards to the south lay a rock corral, measuring approximately eighty-three by thirty-five yards.

Because of the Elliott place's strategic location—midway between the lower JM and upper Quien Sabe—Henry soon made it his headquarters for both ranches. His decision led to the construction of a sizable wooden bunk house about eighty yards west of the rock building. A ten-by-twelve-foot stone milk shed, cooled by water piped from an adjacent windmill, completed the complex.

Henry may have relocated the JM headquarters to the site as early as the summer of 1904. On July 13, the *Texas Stockman-Journal* reported that he had arrived in Midland "from his ranch forty miles south," the approximate distance to the Elliott house.

Henry, who turned thirty in August, also had romance on his mind. He made an extended visit to the Northeast in late summer and fall, and upon his return in early October he announced his engagement to Rosa Wechsler, whom he had met in a lodge in Fabyan, New Hampshire. A New York native, Rosa was Jewish and second-generation Bavarian. Her father had been a pack peddler in Connecticut and Massachusetts before opening a store in New York City.

Presumably, Mayer and Rachel attended the January 3, 1905 wedding, which a rabbi performed at upscale Delmonico's Restaurant in New York while a blizzard raged outside. Henry soon took his bride to San Antonio, where the young woman enjoyed a last few days of city life before Henry would take her to the Quien Sabe to live.

Mayer, at home in two worlds, feared that young Rosa would have difficulty adjusting to his wild western ranch. After all, she had known nothing but the sheltered life of a fashionable New York City debutante for her entire twenty-one years. As he saw her admiring the beautiful gardens and winter flowers along Goliad Street outside his house, his innate kindness showed through.

"Rosa," he said, "you're going out to live on a ranch, and you'll be queen of all you survey for thirty miles in every direction. But if you're not happy there, you come back. You're always welcome here in San Antonio."

Surprisingly, Rosa not only adapted to ranch life, she reveled in it. She and Henry soon located at so-called Section Fourteen in Glasscock County, from which the pair reigned over an immense expanse. Although her trousseau included a specially made habit for riding sidesaddle, she discovered that she had little use for it.

"I never had it on," she recalled. "Henry brought me a pair of Levi's, and in the morning at six o'clock he would bring me a cup of coffee and he would say to me, 'Come on, the old man will catch you.' And when I said, 'Who is the old man?,' he said, 'The sun.' We used to ride and ride and ride all day every day."

By 1905 Mayer's business acumen had made him a million-aire, a rarity in the early twentieth century. Based on a 1906 tally, he had $1.22 million in business holdings that returned an annual profit of more than $116,000. The inventory also reflects the importance of M. Halff & Brother in financing his cattle operation; his 4,170 shares in the merchandising company were worth $417,350 and paid a dividend of $41,600. His largest and oldest ranch, the JM, was valued at $209,603, including live-stock, but delivered a profit of only $12,299.

Mayer's most lucrative spread was the Crouch, which yielded a profit of $43,497 for a $184,121 interest in the land and livestock. His remaining ranches and herds had an aggre-gate worth of $303,247, including $145,975 for the Prince, $106,599 for the Quien Sabe, and $50,673 for the Sabinal.

Mayer also had long since become a Texas land monarch. He had disposed of his Bee County ranch by 1902, but by late 1905 he still owned 179,628 acres and controlled a reported 674,085 to 788,085 acres more. His largest spreads were the Quien Sabe (320,000 to 384,000 acres) and JM (304,000 to 354,000 acres), while the Mallet comprised 92,160 acres, the Schreiner and Halff 66,922 acres, the Prince 52,000 acres, the Sabinal 15,248 acres, and a Bexar County stock farm 640 acres.

His total range land—1,334 to 1,512 square miles, or al-most a million acres—dwarfed the state of Rhode Island and was almost one-third the size of Connecticut. His actual stock num-bers are unknown, but if he grazed only one animal per eigh-teen and a half acres, as he may have done on the arid JM in 1899, he owned as many as 52,000 cattle. Indeed, one source claims that the Halff operation once branded 35,000 calves in a single year, the third-greatest number in the nation.

Soon after the purchase of the Crouch in 1900, Mayer's wealth allowed him to finance a search for a rumored artesian water belt in the Frio and Leona river valleys. Multiple dry holes failed to discourage him, and sometime in 1905 he finally met with success. O. M. Boon, drilling with a Dempster Number 8 Combination machine on Schreiner and Halff land four miles south of Pearsall, struck artesian water at a depth of 1,473 feet.

With a flow of about five hundred gallons a minute, the well opened up enormous possibilities, and soon Mayer and Schreiner were irrigating thirty acres of onions with successful results.

Mayer's discovery of an artesian belt proved a windfall to Frio County residents. Within a few years, dozens of small farms irrigated fields of spinach, onions, cabbage, cauliflower, carrots, beets, and citrus fruits. By the end of the twentieth century, the agricultural character of the county had long since shifted from ranching to farming.

Northeast of Frio in Bexar County, Mayer had stocked a section of land with at least fifty-four cattle and twenty-one horses by early 1905. On February 14 a fire raged for hours across the spread until cowhands from a neighboring ranch brought it under control. The next morning the blaze broke out again and attracted considerable attention in San Antonio, eight miles to the east. "The entire western horizon," said the February 15 *San Antonio Daily Light*, "was wreathed in a dense volume of smoke this morning." The newspaper did not report if Mayer suffered any stock losses.

With good grass blanketing the Pecos and Staked Plains, a winter storm two weeks later claimed only a few old Quien Sabe bulls that Barnes Tullous had transferred to his brother Riley on the Pecos. But it was the passing of another breed—the cattle baron—which the *Texas Stockman Journal* lamented on March 8, portending Mayer's own demise before the end of the year.

"The old order has passed away, taking with it the easy wealth made by the cattle king, who will perhaps never be seen on the plains again," reflected a Midland correspondent. "[He is] like the evening sun declining in the western horizon . . . . Little white cottages are now dotting the plains, the big rancher is giving way to the small stock farmer, and the long-horns are seen here no more. . . . Many men who have hitherto objected to breaking the sod can now be seen making ready for the plow."

One man who understood the ways of the fading pioneer cowman was Theodore Roosevelt, a champion of the cattle industry. Upon venturing to Dakota Territory in the 1880s, he had learned the cow business from top to bottom. "Mingling with cowboys, eating their rough food, sharing their saddles as pillow in the open at night and riding all day in the round-up, this tenderfoot of the East soon became inured to frontier life," recalled the May 31, 1905, *Texas Stockman-Journal.*

As president, Roosevelt's stand on agricultural issues had earned him wide acclaim among cattlemen. On March 23, 1905, the Texas Cattle Raisers' Association adopted a resolution praising him for his "interest . . .in the welfare of the live stock industry of the country, and his fearless actions." Mayer undoubtedly supported Roosevelt, and two weeks later in San Antonio, he had an opportunity to banquet with the renowned leader.

Roosevelt arrived on April 7, and after a day of parades, welcomes, and speeches, he and his entourage gathered for a "swallow-tailed-suit" affair at the Menger Hotel along Alamo Plaza. In addition to Mayer, the guest list of 157 persons included the U.S. secretary of state, former Texas Governor J. D. Sayers, San Antonio's mayor, and Charles Schreiner. Mayer, now in his eleventh hour, obviously had earned the respect of more than just his fellow cattlemen.

In late April and early May, Mayer spent time on the Prince, Crouch, and Sabinal ranches, then he and Rachel prepared to travel abroad for the summer. They were probably at sea on May 29 when tragedy struck in San Antonio. Solomon went to his office as usual that morning but complained of feeling ill and left at noon. In late afternoon he visited his physician and returned to his Goliad Street residence. With his family at his bedside in an upstairs room, Solomon died unexpectedly shortly after 6:00 P.M. The *San Antonio Express* eulogized him the next day as "a promoter of all progression in this city" and a man "highly esteemed for that very fortunate gift of nature—kindness."

It's uncertain when news of Solomon's passing caught up with Mayer and Rachel, but they continued with their journey, arriving in Frankfurt, Germany, on June 16. From there they toured the continent, making stops at Berlin, Lauterbourg, Sweden, Norway, and other sites. On August 23 they embarked for North America from Copenhagen, Denmark, on a 10,000-ton steamer carrying 1,100 passengers and a crew of 200.

Although their son-in-law, Frederick Goldsmith, expressed concerns about Mayer's health in a July 31 letter, Mayer seemed to enjoy this last visit to his homeland. As the steamer passed Newfoundland on September 1, he wrote Alex that "we have had a most delightful time. . . . The eating is good and not one meal has been missed by one of us, and if we do continue, it will be the nicest trip we have had on the ocean."

Despite his early ties to Alsace, Mayer had no regrets about the direction his life had taken since he had first departed Lauterbourg as a mere lad. His choice to emigrate had been the most important decision he had ever made, and this final trip only solidified his emotional commitment to his adopted land of America. "We get enough of all good things," he wrote as the ship approached New York, "and we shall be most happy to get again into our own country."

If the steamer held to schedule, it docked in New York on September 4. Mayer planned to remain in the city for some time, visiting family, seeking eye treatment for Rachel, and checking M. Halff & Brother's New York office, which had a surplus of old items ready to be shipped.

Back home by early November, Mayer engaged in a brief flurry of activity. During the first eight days of the month, he expanded the Quien Sabe and JM with the purchase of 3,040 acres in Midland County, 1,280 acres in Glasscock County, and two tracts totaling more than a section in Upton County. Moreover, it was rush season at M. Halff & Brother, and with Alex in New York on business, Mayer evidently took an active role in company affairs. For an extended period, his employees worked ungrudgingly far into the night, and Mayer was so impressed with their work ethic that he decided to honor them.

On the night of November 8, thirty-four young men gathered in a banquet room to hear Mayer make an impassioned speech about the rewards of integrity and hard work in the world of business. His remarks, spoken from the heart, came as he interrupted yet another glowing toast to his company's accomplishments.

"Talk about yourselves, boys; the firm needs no words," he said with emotion. "And talking about success, I have a message for you each. . . . Don't tell me opportunity is for the rich, boys—it is for you young men who are grappling with the hard facts of life, for you whose soul is in the struggle."

Mayer went on to encourage his employees to live by the ideals so dear to him: integrity, honesty, economy, industry, and politeness. "Character and standing," he said, "can never be taken away from you."

Just before he sat, to face a dusk from which there would be no return, he addressed each young man as a caring father would a son.

"In my life," he said with choked voice, "I have succeeded beyond my expectations. I only wish for each of my boys that he will follow in my footsteps."

Mayer's remarks, published in the *San Antonio Express*, had a powerful impact.

"Probably no sermon preached in San Antonio in many a day has had more appreciation," the *Express* observed. "They were encouraging words spoken by one of the State's foremost business men, especially encouraging because [they were] spoken at a time when many men believe that the adage 'All is fair in love and war,' has been perverted into 'All is fair in money-making.' Mr. Halff is a man of affairs and his enunciation of a guide of conduct carries with it the authority of success."

Three days later, merchant W. B. Adams of Adams & Company in Devine, Texas, penned Mayer a letter.

"I have just had the pleasure of reading in the column of the *Express* your most commendable and excellent address to your employees," said Adams. "Truly your remarks were felicitous and well-timed, and I am sure made the occasion a happy one for your helpers. . . . I now desire to congratulate you for the spirit and kind advice given your young men."

Mayer's "Any Boy May Win" speech, as the *Express* labelled it, served as an appropriate finale to a business life spanning more than half a century and encompassing worlds as diverse as hats and hooves. His health was failing rapidly; he suffered from what was described at the time as an "abscess of the prostate" but which actually may have been prostate cancer. His 1905 tally book contains only two more entries: a November 9 reference to the Crouch ranch, and an undated note reading "Emil Locke to Falls City [Texas]." By early December his illness, exacerbated by a severe cold, was in its final stages.

Resigned to his home and likely to his bed, Mayer found

both solace and distress in family matters as his candle burned ever-shorter. Henry and Rosa were expecting their first child, and the young woman returned to San Antonio for the last days of her pregnancy. In Mayer's home on December 12, Rosa gave birth to a son, and she and Henry honored the infant's dying grandfather by naming him Mayer Henry.

Even so, Mayer may have known loneliness in this midnight of his days. Rachel, who had been at his side for much of the past thirty-nine years, was fifteen hundred miles away in New York, confined to a hospital for reasons perhaps related to her eye condition.

By December 15, if not before, Mayer knew he was dying. At four o'clock that afternoon, he summoned his most trusted friend and confidante, R. A. "Rufe" Moore, and dictated his final wishes:

"In case of death, I wish as pallbearers Capt. [Charles] Schreiner, J. J. Little [one of Mayer's foremen], and R. A. Moore —and to be put away in a plain coffin—and no extravagant service at burial, & to be of short duration."

Even at the end, he clung to his uncomplicated and earthy ways despite his wealth and high social standing. His cowboys would have been proud of him.

On December 20, probably after Mayer had lapsed into a coma, Moore penned Alex a note which reflected his deep feelings for his old friend. "I go back to the ranch this morning," he wrote, "and in as much as I am under promise to your father to give you copy of the enclosed [death instructions], I leave it with you. God knows I wish it was in my power to do something for him."

But there was nothing that anyone could do now, except try to keep him alive until Rachel could return from New York. Stubbornly, Mayer held on, as if sensing Rachel's imminent return and requiring her hand for his final walk into eternity. She arrived on Saturday morning, December 23, and immediately rushed to their home. At 1:40 P.M., as Rachel, their four children, and Rabbi Samuel Marks crowded his bedside, Mayer died peacefully, a man richer in family and friends than in gold and silver.

Two days later, after a funeral service at his home, he was

laid to rest at Temple Beth-El Cemetery with the simple dignity he had wanted.

"We do not pay tribute to his memory because he was wealthy," eulogized Rabbi Marks. "Neither is it because he was a self-made man, for to become such is the privilege of every American. And it is not because of his patriotism, nor his charitable works nor his liberal giving. . . .

"We honor him for his one great quality, towering above all his others—modesty and unostentatious goodness. . . . He descends to his grave with a crown upon his brow and an unsullied name, and he leaves behind the gratitude of the entire community."

But it was an old Quien Sabe cowhand, blessed with a cowboy's simple and honest ways, who delivered perhaps the most fitting tribute to Mayer the man.

"He was a pioneer and built always for the future," wrote Bob Beverly in 1944. "He was the kind that made this the greatest republic the world has ever known. I say 'Peace to his ashes.' I take my hat off to him and his kind."

Appendix

A Sampling of Mayer Halff's Cattle Shipments By Rail Out of Texas

June 1883: 275 head to Chicago.

September 1883: two cars from San Antonio.

January 1885: four cars to San Antonio, five cars to Columbus, five cars to New Orleans.

March 1885: five cars to San Antonio.

May 1885: two train loads to St. Louis.

May 1885: unspecified number to Chicago.

Spring 1887: 1,000 head to A. Mills, Indian Territory; 1,100 head to Indian Territory.

November 1887: 250 head to Chicago.

February 1888: two train loads to A. Mills, Indian Territory.

March 1888: unspecified number to Indian Territory.

June 1888: 20 head to Chicago.

July 1888: 183 head to Chicago.

September 1888: 358 head to Chicago.

October 1888: 236 head to Chicago.

November 1888: 388 head to Chicago.

March 1889: 2,100 head to Kansas.

Spring 1889: 5,000 head to Indian Territory.

August 1889: 146 head to Kansas City.

August 1889: 260 head to Chicago.

September 1889: 270 head to Chicago.

September 1889: 89 head to Kansas City.

September 1889: 418 head to Chicago.

October 1889: 52 head to Kansas City.

October 1889: 106 head to Kansas City.

July 1890: 163 head to Chicago.

July 1890: 192 head to Kansas City.

July 1890: 25 head to Kansas City.

November 1890: 14 head to Kansas City.

September 1891: 320 head to Chicago.

September 1891: 370 head to Chicago.

September 1891: 300 head to Kansas City.

October 1891: 300 head to Chicago.

October 1891: 78 head to Chicago.

October 1891: 450 head to Chicago.

March 1892: 1,800 head to Indian Territory.
July 1892: unspecified number to Chicago.
July 1892: unspecified number to Kansas City.
August 1892: unspecified number to Chicago.
August 1892: unspecified number to Kansas City.
September 1892: unspecified number to Kansas City.
September 1892: unspecified number to Chicago.
October 1892: unspecified number to Kansas City.
April 1893: 2,000 head to Indian Territory.
April 1896: 2,500 head to Indian Territory.
October 1896: unspecified number to Kansas City.
Spring 1897: 2,500 head to Indian Territory.
April 1897: 4,000 head to unspecified location.
June 1897: unspecified number to Kansas City.
Fall 1897: 2,700 head to Kansas City.
November 1897: unspecified number to Kansas City.
March 1898: 40 car loads to Indian Territory.
August 1898: 12 car loads to Kansas.
April 1899: 45 car loads to Indian Territory.
May 1899: 75 car loads to Kansas.
March 1900: 200-plus car loads to Indian Territory.
April 1900: 4,000 head to Kansas City and Indian Territory.
May 1900: 12 car loads to Kansas.
March 1902: 50 car loads to unspecified location.
April 1902: 30 car loads to unspecified location.
September 1902: 344 head to Kansas City.
October 1902: 27 head to Kansas City.
October 1902: 280 head to Chicago.
October 1902: 475 head to Chicago.
October 1902: 175 head to Kansas City.
November 1902: 52 head to Kansas City.
December 1902: 26 head to Kansas City.
January 1903: 84 head to Kansas City.
February 1903: 245 head to Kansas City.
April 1903: 193 head to Kansas City.
May 1903: 378 head to St. Louis.
May 1903: 15 car loads to St. Louis.
May 1903: 333 head to St. Louis.
July 1903: 252 head to Kansas City.
July 1903: 46 head to St. Louis.
November 1903: 105 head to Kansas City.
November 1904: 175 head to Fort Worth.
March 1905: 95 head to Fort Worth.
Summer 1905: train load to unspecified location.

Descendants of Mayer Halff

Compiled by Alex H. Halff and Leonard Robbins

Generation No. 1

1. MAYER[7] HALFF *(HENRY (THE 3RD),[6] SALOMON,[5] SAMUEL,[4] HERTZEL,[3] SALOMON,[2] NAFTALY[1])* was born 06 February 1836 in Lauterbourg, Alsace, France (Source: *acte de naissance*), and died 23 December 1905 in San Antonio, Bexar, Texas (Source: *tombstone*). He married RACHEL HART 1866 in Detroit, Michigan, daughter of ISAAC HART and JULIA COHEN. She was born 10 October 1845 in New Orleans, Louisiana (Source: *tombstone*), and died 17 October 1919 in New York, New York (buried in San Antonio, Bexar, Texas) (Source: *tombstone*).

More About Mayer HALFF:
Cause of Death: Abscess of prostate
Emigration date: U.S. Abt. 1850, San Antonio by 1865, naturalized April 1860
Occupation: General store owner, cattleman, wholesale dry goods businessman, land owner
Other facts: Sold Liberty, Texas, store Abt. August 1860 to Alphonse Dreyfus

More About RACHEL HART:
Other facts: Twin with Morris Hart

Children of MAYER HALFF and RACHEL HART are:
2. i.   HENRIETTA (HENNIE) HALFF, b. 25 September 1867; d. 09 June 1921
3. ii.   ALEXANDER HART HALFF, b. 05 February 1869, Detroit, Michigan; d. 05 November 1942, San Antonio, Bexar, Texas
    iii. ROSA[8] HALFF, b. San Antonio, Bexar, Texas; d. San Antonio, Bexar, Texas
4. iv.   HENRY MAYER HALFF, b. 17 August 1874, San Antonio, Bexar, Texas; d. 21 February 1934, Richardson, Texas

5. v. LILLIE H. HALFF, b. 04 August 1877; d. 01 February 1963, San Antonio, Bexar, Texas

vi. SIDNEY HALFF, b. 23 September 1886, San Antonio, Bexar, Texas (Source: *tombstone*); d. 15 April 1887, San Antonio, Bexar, Texas (Source: *tombstone*)

Generation No. 2

2. HENRIETTA (HENNIE)[8] HALFF (*MAYER,*[7] *HENRY (THE 3RD),*[6] *SALOMON,*[5] *SAMUEL,*[4] *HERTZEL,*[3] *SALOMON,*[2] *NAFTALY*[1]) was born 25 September 1867 and died 09 June 1921. She married FREDERICK GOLDSMITH, who died in 1938.

Children of Henrietta HALFF and FREDERICK GOLDSMITH are:

8. i. RUTH[9] GOLDSMITH, b. 1900
9. ii. WALTER GOLDSMITH
10. iii. LILLIAN GOLDSMITH
11. ivi. GERTRUDE GOLDSMITH, b. 1891; D. 1955

3. ALEXANDER HART[8] HALFF (*MAYER,*[7] *HENRY (THE 3RD),*[6] *SALOMON,*[5] *SAMUEL,*[4] *HERTZEL,*[3] *SALOMON,*[2] *NAFTALY*[1]) was born 05 February 1869 in Detroit, Michigan (Source: *tombstone*), and died 05 November 1942 in San Antonio, Bexar, Texas (Source: *tombstone*). He married ALMA ALTHEA OPPENHEIMER 23 January 1899 in San Antonio, Bexar, Texas, daughter of DANIEL OPPENHEIMER and LOUISA GOLDSTEIN. She was born 21 August 1878 in San Antonio, Bexar, Texas (Source: *tombstone*), and died 11 December 1963 in San Antonio, Bexar, Texas (Source: *tombstone*).

Children of ALEXANDER HALFF and ALMA OPPENHEIMER are:

6. i. HARRY ALEXANDER[9] HALFF, b. 30 October 1899, San Antonio, Bexar, Texas; d. 26 December 1970, San Antonio, Bexar, Texas
7. ii. EVELYN LOUISE HALFF, b. 18 January 1901, San Antonio, Bexar, Texas; d. 31 October 1991, Minneapolis, Minnesota
iii. ROBERT HALFF, b. 01 December 1908, San Antonio, Bexar, Texas

4. HENRY MAYER[8] HALFF (*MAYER,*[7] *HENRY (THE 3RD),*[6] *SALOMON,*[5] *SAMUEL,*[4] *HERTZEL,*[3] *SALOMON,*[2] *NAFTALY*[1]) was born 17 August 1874 in San Antonio, Bexar, Texas (Source: *tombstone*), and died 21 February 1934 in Richardson, Texas (Source: *tombstone*). He married ROSA WECHSLER 03 January 1905 in New York, New York. She was born 07 September 1883 in New York, New York, and died 09 March 1973 in Boston, Massachusetts.

More about HENRY MAYER HALFF:
Other facts: Served in Spanish American War

Children of HENRY HALFF and ROSA WECHSLER are:

12. i.   MAYER HENRY (HAL)[9] HALFF, b. 12 December 1905, San
           Antonio, Bexar, Texas; d. 31 July 1996, Austin, Travis, Texas,
           (buried in San Antonio, Bexar, Texas)
13. ii.   ERNESTINE JULIA HALFF, b. 12 October 1908, New York,
           New York
14. iii. RAY ELIZABETH (BETTY) HALFF, b. 12 June 1911, Midland,
           Texas; d. 28 April 1995, Dallas, Dallas, Texas
15. iv.   ALBERT HENRY HALFF, b. 20 August 1915, Midland, Texas

5. LILLIE H.[8] HALFF (*MAYER,*[7] *HENRY (THE 3RD),*[6] *SALOMON,*[5] *SAMUEL,*[4] *HERTZEL,*[3] *SALOMON,*[2] *NAFTALY*[1]) was born 04 August 1877 (Source: *tombstone*), and died 01 February 1963 in San Antonio, Bexar, Texas (Source: *tombstone*). She married JESSE DANIEL OPPENHEIMER 24 March 1898, son of DANIEL OPPENHEIMER and LOUISA GOLDSTEIN. He was born 13 June 1870 in San Antonio, Bexar, Texas (Source: *tombstone*), and died 30 August 1964 in San Antonio, Bexar, Texas (Source: *tombstone*).

Children of LILLIE HALFF and JESSE OPPENHEIMER are:

16. i.   ELIZABETH[9] OPPENHEIMER, b. 11 June 1904; d. in Portland,
           Oregon
17. ii.   RAE LOUISE OPPENHEIMER, d. Abt. 1993, San Antonio,
           Bexar, Texas
18. iii. HERBERT MAYER OPPENHEIMER, b. 01 June 1911;
           d. 02 June 1989, San Antonio, Bexar, Texas
19. iv.   JESSE HALFF OPPENHEIMER, b. 04 January 1919

Generation No. 3

6. Harry ALEXANDER[9] HALFF (*ALEXANDER HART,*[8] *MAYER,*[7] *HENRY (THE 3RD),*[6] *SALOMON,*[5] *SAMUEL,*[4] *HERTZEL,*[3] *SALOMON,*[2] *NAFTALY*[1]) was born 30 October 1899 in San Antonio, Bexar, Texas (Source: *tombstone*), and died 26 December 1970 in San Antonio, Bexar, Texas (Source: *tombstone*). He married ANYCE BLISS EISEMAN POLLOCK 06 November 1922 in Huntington, West Virginia. She was born 01 November 1900 in Huntington, West Virginia, and died 29 October 1984 in San Antonio, Bexar, Texas.

Children of HARRY HALFF and ANYCE POLLOCK are:

      i.   HOWARD ALEXANDER[10] HALFF, b. 20 December 1925; m.

BETTY FATHEREE MURRAY, 19 February 1994, San Antonio, Bexar, Texas; b. 02 August 1932

ii. THOMAS ALEXANDER HALFF, b. 05 September 1927, San Antonio, Bexar, Texas; d. 01 January 1989, San Antonio, Bexar, Texas; m. CONSUELO DE HEUSCH MUSY, Barcelona, Spain, 15 June 1963, San Antonio, Bexar, Texas; d. 09 September 1979.

More about THOMAS ALEXANDER HALFF:
Cause of death: Multiple sclerosis

20. iii. ALEXANDER HART HALFF II, b. 04 June 1931, San Antonio, Bexar, Texas

7. EVELYN LOUISE[8] HALFF (*ALEXANDER HART,[8] MAYER,[7] HENRY (THE 3RD),[6] SALOMON,[5] SAMUEL,[4] HERTZEL,[3] SALOMON,[2] NAFTALY[1]*) was born 18 January 1901 in San Antonio, Bexar, Texas, and died 31 October 1991 in Minneapolis, Minnesota. She married EDMUND RUBEN. He was born 05 August 1898, in Cleveland, Ohio, and died 05 October 1991 in Minneapolis, Minnesota.

Children of EVELYN HALFF and EDMUND RUBEN are:
21. i. NANCY[10] RUBEN, b. 13 March 1924 in Minneapolis, Minnesota
22. ii. THOMAS RUBEN, b. 15 June 1930, in New York, New York

8. RUTH[9] GOLDSMITH (*HENRIETTA (HENNIE)[8] HALFF, MAYER,[7] HENRY (THE 3RD),[6] SALOMON,[5] SAMUEL,[4] HERTZEL,[3] SALOMON,[2] NAFTALY[1]*) was born 1900. She married HILBERT BAIR.

Child of RUTH GOLDSMITH and HILBERT BAIR is:
23. i. HILDEGARDE[10] BAIR

9. WALTER[9] GOLDSMITH (*HENRIETTA (HENNIE)[8] HALFF, MAYER,[7] HENRY (THE 3RD),[6] SALOMON,[5] SAMUEL,[4] HERTZEL,[3] SALOMON,[2] NAFTALY[1]*). He married ROSETTA C.

Children of WALTER GOLDSMITH and ROSETTA C. are:
i. ROBERT[10] GOLDSMITH, m. LYNN
24. i. KATE A. GOLDSMITH

10. LILLIAN[9] GOLDSMITH (*HENRIETTA (HENNIE)[8] HALFF, MAYER,[7] HENRY (THE 3RD),[6] SALOMON,[5] SAMUEL,[4] HERTZEL,[3] SALOMON,[2] NAFTALY[1]*). She married ROWLAND D. GOODMAN.

Children of LILLIAN GOLDSMITH and ROWLAND GOODMAN are:

25.   i.   ROWLAND D.[10] GOODMAN II, b. 10 October 1917

     ii.   FREDERICK GOODMAN, b. 16 October 1919; d. 1944

26.   iii.   FRANK GOODMAN, b. 30 January 1923

11. GERTRUDE[9] GOLDSMITH (*HENRIETTA (HENNIE)*[8] *HALFF, MAYER,*[7] *HENRY (THE 3RD),*[6] *SALOMON,*[5] *SAMUEL,*[4] *HERTZEL,*[3] *SALOMON,*[2] *NAFTALY*[1]) was born 1891 and died 1955. She married BERNARD A. ROSENBLATT.

Children of GERTRUDE GOLDSMITH and BERNARD ROSENBLATT are:

27.   i.   JONATHAN[10] ROSENBLATT, b. 1919; d. August 1994

28.   ii.   DAVID ROSENBLATT, b. 1918

12. MAYER HENRY (HAL)[9] HALFF (*HENRY MAYER,*[8] *MAYER,*[7] *HENRY (THE 3RD),*[6] *SALOMON,*[5] *SAMUEL,*[4] *HERTZEL,*[3] *SALOMON,*[2] *NAFTALY*[1]) was born 12 December 1905 in San Antonio, Bexar, Texas (Source: *tombstone*), and died 31 July 1996 in Austin, Travis, Texas, (buried in San Antonio, Bexar, Texas) (Source: *tombstone*). He married (1) MAUREENE VICTORIA VAN POOLE 05 August 1933 in Durant, Oklahoma. She was born 06 April 1913, and died 06 September 1990 in Dallas, Dallas, Texas. He married (2) WINNIE VAN KEULIN 26 February 1969 in San Antonio, Bexar, Texas. She was born 16 September 1914 in Grand Rapids, Michigan. He married (3) IONE (SISSY) BISHOP.

Children of MAYER HALFF and MAUREENE VAN POOLE are:

29.   i.   EVELYN JULIA[10] HALFF, b. 16 April 1934, El Paso, Texas

30.   ii.   HENRY RICHARD HALFF, b. 10 June 1935, El Paso, Texas

13. ERNESTINE JULIA[9] HALFF (*HENRY MAYER,*[8] *MAYER,*[7] *HENRY (THE 3RD),*[6] *SALOMON,*[5] *SAMUEL,*[4] *HERTZEL,*[3] *SALOMON,*[2] *NAFTALY*[1]) was born 12 October 1908 in New York, New York. She married NORMAN FREEMAN 15 June 1931 in Boston, Massachusetts. He was born 22 July 1900, and died 15 May 1986 in Dallas, Dallas, Texas.

Children of ERNESTINE HALFF and NORMAN FREEMAN are:

31.   i.   PATRICIA[10] FREEMAN, b. 29 May 1932

     ii.   RONNIE FREEMAN, b. 22 October 1941; d. 25 April 1954, Dallas, Dallas, Texas

More about RONNIE FREEMAN:

Cause of death: Meningitis following head injury

14. RAY ELIZABETH (BETTY)[9] HALFF (*HENRY MAYER,*[8] *MAYER,*[7] *HENRY*

(THE 3RD),[6] SALOMON,[5] SAMUEL,[4] HERTZEL,[3] SALOMON,[2] NAFTALY[1]) was born 12 June 1911 in Midland, Texas, and died 28 April 1995 in Dallas, Dallas, Texas. She married (1) MARTIN ZINN, JR. on 09 November 1935. He was born 04 December 1911, and died 26 August 1982. She married (2) GEORGE WILLIAM LLEWELLYN 13 November 1947. He was born 06 June. She married (3) BOB O'LEARY after 1949.

Children of RAY HALFF and MARTIN ZINN are:

32.   i.   MARTIN (MARTY)[10] ZINN III, b. 10 June 1941, Lake Charles, Louisiana
     ii.   HENRY HAROLD ZINN, b. 06 August 1943
33.   iii.   MARY ELIZABETH ZINN, b. 08 April 1945, Lake Charles, Louisiana

15. ALBERT HENRY[9] HALFF *(HENRY MAYER,[8] MAYER,[7] HENRY (THE 3RD),[6] SALOMON,[5] SAMUEL,[4] HERTZEL,[3] SALOMON,[2] NAFTALY[1])* was born 20 August 1915 in Midland, Texas. He married LEE CATHERINE BENSON 24 August 1940 in Chicago, Illinois. She was born 13 October 1914 in Knoxville, Iowa.

Children of ALBERT HALFF and LEE BENSON are:

34.   i.   HENRY MAYER (LEFTY)[10] HALFF, b. 26 November 1942
     ii.   ALBERT LEE (BRO) HALFF, b. 29 March 1944 in Dallas, Texas
     iii.   HALFF, b. 09 December 1946; d. 10 December 1946 in Myrtle Beach, South Carolina

16. ELIZABETH[9] OPPENHEIMER *(LILLIE H.[8] HALFF, MAYER,[7] HENRY (THE 3RD),[6] SALOMON,[5] SAMUEL,[4] HERTZEL,[3] SALOMON,[2] NAFTALY[1])*. She was born 11 June 1904 in San Antonio, Texas. She married LEE DANIEL J. COHN 10 May 1927. He was born 08 November 1902 and died 29 July 1992 in Portland, Oregon.

Children of ELIZABETH OPPENHEIMER and DANIEL COHN are:

35.   i.   PAUL[10] COHN. b. 14 July 1936
36.   ii.   JOANN LOUISE COHN, b. 22 January 1931, m. ROBERT CAZDEN

17. RAE LOUISE[9] OPPENHEIMER *(LILLIE H.[8] HALFF, MAYER,[7] HENRY (THE 3RD),[6] SALOMON,[5] SAMUEL,[4] HERTZEL,[3] SALOMON,[2] NAFTALY[1])* born Abt. 1910 in Texas, died 14 August 1993 in San Antonio, Bexar, Texas. She married ALFRED BURNHAM.

Children of RAE OPPENHEIMER and ALFRED BURNHAM are:

     i.   DORIS (ALICIA)[10] BURNHAM, m. RICHARD WINSLOW
37.   ii.   ELIZABETH BURNHAM

18. HERBERT MAYER[9] OPPENHEIMER (*LILLIE H.*[8] *HALFF, MAYER,*[7] *HENRY (THE 3RD),*[6] *SALOMON,*[5] *SAMUEL,*[4] *HERTZEL,*[3] *SALOMON,*[2] *NAFTALY*[1]) was born 01 June 1911 (Source: *tombstone*), and died 02 June 1989 in San Antonio, Bexar, Texas (Source: *tombstone*). He married MARIAN R. BLOCK 21 September 1937.

Child of HERBERT OPPENHEIMER and MARIAN BLOCK is:
38.   i.   JOHN M.[10] OPPENHEIMER, b. 24 May 1942

19. JESSE HALFF[9] OPPENHEIMER (*LILLIE H.*[8] *HALFF, MAYER,*[7] *HENRY (THE 3RD),*[6] *SALOMON,*[5] *SAMUEL,*[4] *HERTZEL,*[3] *SALOMON,*[2] *NAFTALY*[1]) was born 04 January 1919. He married SUSAN R. ROSENTHAL 18 July 1946 in Boston, Massachusetts. She was born 31 March 1924.

More about JESSE HALFF OPPENHEIMER:
Occupation: Banker, lawyer

Children of JESSE OPPENHEIMER and SUSAN ROSENTHAL are:
39.   i.   J. DAVID OPPENHEIMER, b. 31 July 1949
     ii.   JEAN LOUISE[10] OPPENHEIMER, b. 26 April 1951
40.   iii.   BARBARA SUE OPPENHEIMER, b. 10 December 1959; m. JOHN COHN, b. 18 January 1963

Generation No. 4

20. ALEXANDER HART[10] HALFF II (*HARRY ALEXANDER,*[9] *ALEXANDER HART,*[8] *MAYER,*[7] *HENRY (THE 3RD),*[6] *SALOMON,*[5] *SAMUEL,*[4] *HERTZEL,*[3] *SALOMON,*[2] *NAFTALY*[1]) was born 04 June 1931 in San Antonio, Bexar, Texas. He married SALLY MARCUS MESSER 03 December 1953 in Valejo, California. She was born 04 April 1931 in Omaha, Douglas, Nebraska.

Children of ALEXANDER HALFF and SALLY MESSER are:
41.   i.   GLENN ALEXANDER[11] HALFF, b. 23 January 1957; San Antonio, Bexar, Texas
42.   ii.   HARRY ALEXANDER HALFF II, b. 14 August 1958, San Antonio, Bexar, Texas
     iii.   James ALEXANDER HALFF, b. 25 March 1963

21. NANCY[10] RUBEN (*EVELYN LOUISE*[9] *HALFF, ALEXANDER HART,*[8] *MAYER,*[7] *HENRY (THE 3RD),*[6] *SALOMON,*[5] *SAMUEL,*[4] *HERTZEL,*[3] *SALOMON,*[2] *NAFTALY*[1]) was born 13 March 1924. She married LARRY BENTSON. He was born 17 June 1921.

Children of NANCY RUBEN and LARRY BENTSON are:
43. i. LAURIE[11] BENTSON, b. 04 March 1947
     ii. JAN BENTSON, b. 1950; d. 2000

22. THOMAS[10] RUBEN (*EVELYN LOUISE[9] HALFF, ALEXANDER HART,[8] MAYER,[7] HENRY (THE 3RD),[6] SALOMON,[5] SAMUEL,[4] HERTZEL,[3] SALOMON,[2] NAFTALY[1]*) was born 15 June 1930. He married (1) MARGARET STOVER 13 June 1953. He married (2) Barbara HOLT 11 July 1976. She was born 14 April 1947 in Chicago, Cook, Illinois.

Children of THOMAS RUBEN and MARGARET STOVER are:
  i. DEBBIE (VOLA) RUBEN, b. 27 September 1954, Fort Ord, California
  ii. EDMUND (EDDIE) WILLARD RUBEN, b. 12 September 1955, Minneapolis, Minnesota

23. HILDEGARDE[10] BAIR (*RUTH[9] GOLDSMITH, HENRIETTA (HENNIE)[8] HALFF, MAYER,[7] HENRY (THE 3RD),[6] SALOMON,[5] SAMUEL,[4] HERTZEL,[3] SALOMON,[2] NAFTALY[1]*). She married ARTHUR D. LEWIS.

Children of HILDEGARDE BAIR and ARTHUR LEWIS are:
  i. GREGORY[11] LEWIS
  ii. KIMBERLY LEWIS

24. KATE A.[10] GOLDSMITH (*WALTER,[9] HENRIETTA (HENNIE)[8] HALFF, MAYER,[7] HENRY (THE 3RD),[6] SALOMON,[5] SAMUEL,[4] HERTZEL,[3] SALOMON,[2] NAFTALY[1]*). She married LEONARD F. PEYSER.

Children of KATE GOLDSMITH and LEONARD PEYSER are:
  i. MICHAEL[11] PEYSER
  ii. PATRICIA ANN PEYSER, m. RUSSEL MOLL

25. ROWLAND D.[10] GOODMAN II (*LILLIAN[9] GOLDSMITH, HENRIETTA (HENNIE)[8] HALFF, MAYER,[7] HENRY (THE 3RD),[6] SALOMON,[5] SAMUEL,[4] HERTZEL,[3] SALOMON,[2] NAFTALY[1]*) was born 08 October 1917. He married JUDY (RUTH). She was born 03 March 1928.

Children of ROWLAND GOODMAN and JUDY (RUTH) are:
44. i. ROWLAND D.[11] GOODMAN III
45. ii. JOYCE GOODMAN
     iii. Gwen GOODMAN, m. CHRISTOPHER LOWE, 22 June 1991
     iv. PETER GOODMAN

26. FRANK[10] GOODMAN (*LILLIAN[9] GOLDSMITH, HENRIETTA (HEN-*

NIE)[8] *HALFF, MAYER,*[7] *HENRY (THE 3RD),*[6] *SALOMON,*[5] *SAMUEL,*[4] *HERT-ZEL,*[3] *SALOMON,*[2] *NAFTALY*[1]) was born 30 January 1923. He married NAOMI. She died 12 April 1988.

Children of FRANK GOODMAN and NAOMI are:
- i. DENNIS[11] GOODMAN
- ii. GARY GOODMAN
- iii. Gail GOODMAN
- iv. ANDREA GOODMAN

27. JONATHAN[10] ROSENBLATT (*GERTRUDE*[9] *GOLDSMITH, HENRIETTA (HENNIE)*[8] *HALFF, MAYER,*[7] *HENRY (THE 3RD),*[6] *SALOMON,*[5] *SAMUEL,*[4] *HERTZEL,*[3] *SALOMON,*[2] *NAFTALY*[1]) was born in 1919 and died August 1994. He married DOROTHY. She died in 1994.

Child of JONATHAN ROSENBLATT and DOROTHY is:
- i. DANIEL[11] ROSENBLATT, b. 1946

28. DAVID[10] ROSENBLATT (*GERTRUDE*[9] *GOLDSMITH, HENRIETTA (HENNIE)*[8] *HALFF, MAYER,*[7] *HENRY (THE 3RD),*[6] *SALOMON,*[5] *SAMUEL,*[4] *HERTZEL,*[3] *SALOMON,*[2] *NAFTALY*[1]) was born in 1918. He married CAROL B.

Children of DAVID ROSENBLATT and CAROL B. are:
- i. JOSIAH[11] ROSENBLATT, b. 1947
- ii. LIONALA ROSENBLATT, b. 1943
- 46. iii. NATHANIEL ROSENBLATT, b. 1955
- 47. iv. SARAH TESS ROSENBLATT

29. EVELYN JULIA[10] HALFF (*MAYER HENRY (HAL),*[9] *HENRY MAYER,*[8] *MAYER,*[7] *HENRY (THE 3RD),*[6] *SALOMON,*[5] *SAMUEL,*[4] *HERTZEL,*[3] *SALOMON,*[2] *NAFTALY*[1]) was born 16 April 1934 in El Paso, Texas. She married CHARLES ROBERT KEITH 27 November 1957 in Dallas, Dallas, Texas.

Children of EVELYN HALFF and CHARLES KEITH are:
- 48. i. MERRIE LYNN[11] KEITH, b. 08 August 1958
- 49. ii. JULIA ELIZABETH KEITH, b. 08 September 1959

30. HENRY RICHARD[10] HALFF (*MAYER HENRY (HAL),*[9] *HENRY MAYER,*[8] *MAYER,*[7] *HENRY (THE 3RD),*[6] *SALOMON,*[5] *SAMUEL,*[4] *HERTZEL,*[3] *SALOMON,*[2] *NAFTALY*[1]) was born 10 June 1935 in El Paso, Texas. He married SHIRLEY JANETTE McCORKLE 04 May 1963 in Collinsville, Texas. She was born 04 December 1939.

Children of H<small>ENRY</small> HALFF and S<small>HIRLEY</small> M<small>C</small>CORKLE are:
50.   i.   P<small>HILLIP</small> H<small>ENRY</small>[11] H<small>ALFF</small>, b. 30 May 1964
      ii.   S<small>USAN</small> V<small>ICTORIA</small> H<small>ALFF</small>, b. 09 March 1967; m.
          C<small>RAIG</small> C<small>OPELAND</small>, b. 30 October 1964, m. 12 May 1992,
          Denton, Texas
51.   iii.   S<small>ANDRA</small> J<small>ANETTE</small> H<small>ALFF</small>, b. 03 April 1969
      iv.   S<small>HARON</small> L<small>UCILLE</small> H<small>ALFF</small>, b. 20 April 1970, Dallas, Dallas,
          Texas

31. P<small>ATRICIA</small>[10] FREEMAN (*E<small>RNESTINE</small> J<small>ULIA</small>*[9] *HALFF, H<small>ENRY</small> M<small>AYER</small>,*[8] *M<small>AYER</small>,*[7] *H<small>ENRY</small> (T<small>HE</small> 3<small>RD</small>),*[6] *S<small>ALOMON</small>,*[5] *S<small>AMUEL</small>,*[4] *H<small>ERTZEL</small>,*[3] *S<small>ALOMON</small>,*[2] *N<small>AFTALY</small>*[1]) was born 28 May 1932. She married W<small>ILLIAM</small> LEHRBURGER 08 June 1952 in Brookline, Massachusetts.

Children of P<small>ATRICIA</small> FREEMAN and W<small>ILLIAM</small> LEHRBURGER are:
52.   i.   J<small>AMES</small> F<small>REEMAN</small>[11] L<small>EHRBURGER</small>, b. 15 October 1954
      ii.   A<small>NN</small> L<small>EHRBURGER</small>, b. 14 December 1957; m.
          P<small>AUL</small> V<small>ILLANI</small>. They divorced.
      iii.   R<small>OBERT</small> W<small>ILLIAM</small> L<small>EHRBURGER</small>, b. 26 March 1963

32. M<small>ARTIN</small> (M<small>ARTY</small>)[10] ZINN III (*R<small>AY</small> E<small>LIZABETH</small> (B<small>ETTY</small>)*[9] *HALFF, H<small>ENRY</small> M<small>AYER</small>,*[8] *M<small>AYER</small>,*[7] *H<small>ENRY</small> (T<small>HE</small> 3<small>RD</small>),*[6] *S<small>ALOMON</small>,*[5] *S<small>AMUEL</small>,*[4] *H<small>ERT</small>-Z<small>EL</small>,*[3] *S<small>ALOMON</small>,*[2] *N<small>AFTALY</small>*[1]) was born 10 June 1941 in Lake Charles, Louisiana. He married J<small>UDY</small> DEUTCH. She was born 10 September 1944. They divorced.

Children of M<small>ARTIN</small> ZINN and J<small>UDY</small> DEUTCH are:
      i.   K<small>ENNETH</small> M<small>ARTIN</small>[11] Z<small>INN</small>
      ii.   J<small>ENNIFER</small> R<small>OSE</small> Z<small>INN</small>
      iii.   D<small>ANIEL</small> N<small>ATHAN</small> Z<small>INN</small>, b. 02 April

33. M<small>ARY</small> E<small>LIZABETH</small>[10] ZINN (*R<small>AY</small> E<small>LIZABETH</small> (B<small>ETTY</small>)*[9] *HALFF, H<small>ENRY</small> M<small>AYER</small>,*[8] *M<small>AYER</small>,*[7] *H<small>ENRY</small> (T<small>HE</small> 3<small>RD</small>),*[6] *S<small>ALOMON</small>,*[5] *S<small>AMUEL</small>,*[4] *H<small>ERTZEL</small>,*[3] *S<small>ALOMON</small>,*[2] *N<small>AFTALY</small>*[1]) was born 08 April 1945 in Lake Charles, Louisiana. She married J<small>OSEPH</small> D<small>ENZIL</small> STEWART, Jr., 12 August. He was born 14 January 1945 in Springfield, Ohio. They divorced.

Children of M<small>ARY</small> ZINN and J<small>OSEPH</small> STEWART are:
      i.   A<small>NN</small> E<small>LIZABETH</small>[11] S<small>TEWART</small>, b. 17 March 1970
      ii.   J<small>ULIE</small> L<small>YNN</small> S<small>TEWART</small>, b. 17 April 1972
      iii.   M<small>ICHAEL</small> B<small>ENJAMIN</small> S<small>TEWART</small>, b. 14 November 1984

34. H<small>ENRY</small> M<small>AYER</small> (L<small>EFTY</small>)[10] HALFF (*A<small>LBERT</small> H<small>ENRY</small>,*[9] *H<small>ENRY</small> M<small>AYER</small>,*[8] *M<small>AYER</small>,*[7] *H<small>ENRY</small> (T<small>HE</small> 3<small>RD</small>),*[6] *S<small>ALOMON</small>,*[5] *S<small>AMUEL</small>,*[4] *H<small>ERTZEL</small>,*[3] *S<small>ALOMON</small>,*[2]

NAFTALY[1]) was born 26 November 1942 in McAlester, Oklahoma. He married (1) NANCY FAHLBERG 24 June 1968 in Los Angeles, California. She was born in Los Angeles, 25 May 1941. He married (2) JEAN WATSON 12 April 1997 in San Antonio, Bexar, Texas. She was born in London, England, 27 July 1938.

Child of HENRY HALFF and NANCY is:
    i.   LAWRENCE[11] HALFF, b. 08 October 1970

35. PAUL[10] COHN (*ELIZABETH*[9] *OPPENHEIMER, LILLIE H.*[8] *HALFF, MAYER,*[7] *HENRY (THE 3RD),*[6] *SALOMON,*[5] *SAMUEL,*[4] *HERTZEL,*[3] *SALOMON,*[2] *NAFTALY*[1]).

Children of PAUL COHN are:
53.   i.   LISA[11] COHN, b. 23 July 1962
    ii. JILL COHN, b. 02 June 1964
   iii. MELINDA COHN, b. 11 November 1960; m. DAVID LUCAS; b. 21 July 1955

36. JOANN LOUISE[10] COHN (*ELIZABETH*[9] *OPPENHEIMER, LILLIE W.*[8] *HALFF, MAYER,*[7] *HENRY (THE 3RD),*[6] *SALOMON,*[5] *SAMUEL,*[4] *HERTZEL,*[3] *SALOMON,*[2] *NAFTALY*[1]) was born 22 January 1931. She married ROBERT CAZDEN 16 January 1958. He was born 29 August 1930.

Children of JOANN COHN and ROBERT CAZDEN are:
    i.   DAVID BENJAMIN[11] CAZDEN, b. 06 July 1958
    ii. RICHARD JESSE CAZDEN, b. 11 July 1961; d. 15 July 1961
   iii. ROGER PHILLIP CAZDEN, b. 10 September 1963
   iv. EUGENE LAEL CAZDEN, b. 20 July 1966

37. ELIZABETH[10] BURNHAM (*RAE LOUISE*[9] *OPPENHEIMER, LILLIE H.*[8] *HALFF, MAYER,*[7] *HENRY (THE 3RD),*[6] *SALOMON,*[5] *SAMUEL,*[4] *HERTZEL,*[3] *SALOMON,*[2] *NAFTALY*[1]). She married STEVEN NEVEN.

Children of ELIZABETH BURNHAM and STEVEN NEVEN are:
    i.   NEVEN[11]
    ii. NEVEN

38. JOHN M.[10] OPPENHEIMER (*HERBERT MAYER,*[9] *LILLIE H.*[8] *HALFF, MAYER,*[7] *HENRY (THE 3RD),*[6] *SALOMON,*[5] *SAMUEL,*[4] *HERTZEL,*[3] *SALOMON,*[2] *NAFTALY*[1]) was born 24 May 1942. He married (1) ERICA ERICH April 1966. She died 11 June 1981. He married (2) KATHI MARCHER in San Antonio.

Children of JOHN OPPENHEIMER and ERICA ERICH are:

　　ii.　LAURA[11] OPPENHEIMER, b. 04 March 1967; m. ROBERT
　　　　 CHARLES HARRIS, 30 September 1995
　　iii.　MARIAN OPPENHEIMER, b. 08 March 1969

Child of JOHN OPPENHEIMER and KATHI MARCHER is:

　　i.　ERICA[11] OPPENHEIMER, b. 11 November 1983

39. J. DAVID[10] OPPENHEIMER (*JESSE HALFF,*[9] *LILLIE H.*[8] *HALFF,*
MAYER,[7] *HENRY (THE 3RD),*[6] *SALOMON,*[5] *SAMUEL,*[4] *HERTZEL,*[3] *SALOMON,*[2]
NAFTALY[1]). He married HARRIET SAPAROW 21 November 1981. She was
born 21 October 1954.

Children of J. OPPENHEIMER and HARRIET SAPAROW are:

　　i.　REBECCA[11] OPPENHEIMER, b. 22 February 1983
　　ii.　DANIEL OPPENHEIMER, b. 31 January 1985
　　iii.　JACOB OPPENHEIMER, b. 05 September 1988

40. BARBARA SUE[10] OPPENHEIMER (*JESSE HALFF,*[9] *LILLIE H.*[8] *HALFF,*
MAYER,[7] *HENRY (THE 3RD),*[6] *SALOMON,*[5] *SAMUEL,*[4] *HERTZEL,*[3] *SALOMON,*[2]
NAFTALY[1]) was born 10 December 1959. She married JOHN COHN, who
was born 18 January 1963.

Child of BARBARA SUE OPPENHEIMER and JOHN COHN is:

　　i.　HARRISON COHN, b. 04 June 1999

Generation No. 5

41. GLENN ALEXANDER[11] HALFF (*ALEXANDER HART,*[10] *HARRY ALEXAN-*
DER,[9] *ALEXANDER HART,*[8] *MAYER,*[7] *HENRY (THE 3RD),*[6] *SALOMON,*[5] *SAM-*
UEL,[4] *HERTZEL,*[3] *SALOMON,*[2] *NAFTALY*[1]) was born 23 January 1957 in San
Antonio, Bexar, Texas. He married MINDI HELENE ALTERMAN, daugh-
ter of WILLIAM ALTERMAN and YVETTE LEVINE, 24 March 1984 in San
Antonio, Bexar, Texas. She was born 01 March 1960, in San Antonio,
Texas.

More about GLENN ALEXANDER HALFF:
Occupation: Transplant surgeon

Children of GLENN HALFF and MINDI ALTERMAN are:

　　i.　ALLISON ALTERMAN[12] HALFF, b. 02 February 1987,
　　　　New York, New York
　　ii.　JESSE ALTERMAN HALFF, b. 13 July 1989, Pittsburgh,
　　　　Allegheny, Pennsylvania

iii.   MARGARET (Marghi) ALTERMAN, HALFF, b. 30 June 1992, New York, New York

42. HARRY ALEXANDER[11] HALFF II (*ALEXANDER HART,*[10] *HARRY ALEXANDER,*[9] *ALEXANDER HART,*[8] *MAYER,*[7] *HENRY (THE 3RD),*[6] *SALOMON,*[5] *SAMUEL,*[4] *HERTZEL,*[3] *SALOMON,*[2] *NAFTALY*[1]) was born 14 August 1958 in San Antonio, Bexar, Texas. He married ELIZABETH (LISA) ROMANO, daughter of Raymond ROMANO and Lou, 02 September 1988 in San Antonio, Bexar, Texas. She was born 03 January 1959.

More about HARRY HALFF:
Occupation: Art gallery owner.

Children of HARRY HALFF and ELIZABETH ROMANO are:
i.   TRAVIS A.[12] HALFF, b. 21 January 1991, San Antonio, Bexar, Texas
ii.   CODY A. HALFF, b. 22 June 1992, San Antonio, Bexar, Texas

43. LAURIE[11] BENTSON (*NANCY*[10] *RUBEN,* *EVELYN LOUISE*[9] *HALFF,* *ALEXANDER HART,*[8] *MAYER,*[7] *HENRY (THE 3RD),*[6] *SALOMON,*[5] *SAMUEL,*[4] *HERTZEL,*[3] *SALOMON,*[2] *NAFTALY*[1]) was born 04 March 1947. She married BILL KAUTH. He was born 31 May 1952.

Child of LAURIE BENTSON and BILL KAUTH is:
i.   KIMBERLY[12] KAUTH, b. 22 January 1987

44. ROWLAND D.[11] GOODMAN III (*ROWLAND D.,*[10] *LILLIAN*[9] *GOLDSMITH,* *HENRIETTA (HENNIE)*[8] *HALFF,* *MAYER,*[7] *HENRY (THE 3RD),*[6] *SALOMON,*[5] *SAMUEL,*[4] *HERTZEL,*[3] *SALOMON,*[2] *NAFTALY*[1]). He married ROSLYN GALLO.

Child of ROWLAND GOODMAN and ROSLYN GALLO is:
i.   ROWLAND D.[12] GOODMAN IV, b. 03 September 1991

45. JOYCE[11] GOODMAN (*ROWLAND D.,*[10] *LILLIAN*[9] *GOLDSMITH,* *HENRIETTA (HENNIE)*[8] *HALFF,* *MAYER,*[7] *HENRY (THE 3RD),*[6] *SALOMON,*[5] *SAMUEL,*[4] *HERTZEL,*[3] *SALOMON,*[2] *NAFTALY*[1]). She married ALBERT SIGNORELLA in 1987.

Children of JOYCE GOODMAN and ALBERT SIGNORELLA are:
i.   MICHAEL[12] GOODMAN, b. 30 December 1987
ii.   JENNIFER JOYCE GOODMAN, b. 15 September 1989
iii. JULIA DAVIS GOODMAN, b. 15 September 1989

46. NATHANIEL[11] ROSENBLATT (*DAVID,*[10] *GERTRUDE*[9] *GOLDSMITH,* *HENRIETTA (HENNIE)*[8] *HALFF,* *MAYER,*[7] *HENRY (THE 3RD),*[6] *SALOMON,*[5]

SAMUEL,[4] HERTZEL,[3] SALOMON,[2] NAFTALY[1]) was born 1955. He married LESLIE A. NICHOLS.

Children of NATHANIEL ROSENBLATT and LESLIE NICHOLS are:

    i.   JESSICA[12] ROSENBLATT, b. 04 September 1982
   ii.   ALEXANDER ROSENBLATT, b. 21 January 1985
  iii.   ELIAS G. ROSENBLATT, b. 29 August 1988

47. SARAH TESS[11] ROSENBLATT (*DAVID,[10] GERTRUDE[9] GOLDSMITH, HENRIETTA (HENNIE)[8] HALFF, MAYER,[7] HENRY (THE 3RD),[6] SALOMON,[5] SAMUEL,[4] HERTZEL,[3] SALOMON,[2] NAFTALY[1]*). She married JACKSON. He was born 1945.

Child of SARAH ROSENBLATT and JACKSON is:

    i.   BENJAMIN D.[12] JACKSON, b. 03 February 1984

48. MERRIE LYNN[11] KEITH (*EVELYN JULIA[10] HALFF, MAYER HENRY (HAL),[9] HENRY MAYER,[8] MAYER,[7] HENRY (THE 3RD),[6] SALOMON,[5] SAMUEL,[4] HERTZEL,[3] SALOMON,[2] NAFTALY[1]*) was born 08 August 1958. She married SCOTT GOUGEN.

Children of MERRIE KEITH and SCOTT GOUGEN are:

    i.   JOSHUA[12] GOUGEN, b. 23 May 1972
   ii.   JONATHAN GOUGEN, b. 22 February 1983

49. JULIA ELIZABETH[11] KEITH (*EVELYN JULIA[10] HALFF, MAYER HENRY (HAL),[9] HENRY MAYER,[8] MAYER,[7] HENRY (THE 3RD),[6] SALOMON,[5] SAMUEL,[4] HERTZEL,[3] SALOMON,[2] NAFTALY[1]*) was born 08 September 1959. She married JOHN SCHMALSTEIG III. He was born 16 February 1954.

Child of JULIA KEITH and JOHN SCHMALSTEIG is:

    i.   JOHN[12] SCHMALSTEIG IV, b. 19 May 1992

50. PHILLIP HENRY[11] HALFF (*HENRY RICHARD,[10] MAYER HENRY (HAL),[9] HENRY MAYER,[8] MAYER,[7] HENRY (THE 3RD),[6] SALOMON,[5] SAMUEL,[4] HERTZEL,[3] SALOMON,[2] NAFTALY[1]*) was born 30 May 1964. He married THERESA WILLIAMS.

Child of PHILLIP HALFF and THERESA WILLIAMS is:

    i.   PHILLIP HENRY[12] HALFF II, b. 12 June 1991

51. SANDRA JANETTE[11] HALFF (*HENRY RICHARD,[10] MAYER HENRY (HAL),[9] HENRY MAYER,[8] MAYER,[7] HENRY (THE 3RD),[6] SALOMON,[5] SAMUEL,[4] HERTZEL,[3] SALOMON,[2] NAFTALY[1]*) was born 03 April 1969. She married SHAWN ETHERIDGE.

Children of SANDRA HALFF and SHAWN ETHERIDGE are:

    i.   NICHOLE[12] ETHERIDGE, b. 25 January 1988
   ii.   JESSICA M. ETHERIDGE, b. 26 May 1992

52. JAMES FREEMAN[11] LEHRBURGER (*PATRICIA[10] FREEMAN, ERNESTINE JULIA[9] HALFF, HENRY MAYER,[8] MAYER,[7] HENRY (THE 3RD),[6] SALOMON,[5] SAMUEL,[4] HERTZEL,[3] SALOMON,[2] NAFTALY[1]*) was born 15 October 1954. He married KAREN MEIDENBAUER.

Children of JAMES LEHRBURGER and KAREN MEIDENBAUER are:
    i.   STEVEN J.[12] LEHRBURGER, b. 03 April 1985
   ii.   AMY E. LEHRBURGER, b. 01 July 1988

53. LISA[11] COHN (*PAUL,[10] ELIZABETH[9] OPPENHEIMER, LILLIE H.[8] HALFF, MAYER,[7] HENRY (THE 3RD),[6] SALOMON,[5] SAMUEL,[4] HERTZEL,[3] SALOMON,[2] NAFTALY[1]*) was born 23 July 1962. She married JEFF SHAFFER. He was born 08 November 1960.

Child of LISA COHN and JEFF SHAFFER is:
    i.   EUGENE[12] SHAFFER, b. 21 July 1995

Bibliography

AUTHOR'S NOTE: Although detailed documentation was not required in this work, if a researcher has questions regarding specific sources, he may direct inquiries to me in care of the publisher.

My most important source was the Halff Family Collection, in the possession of Alex H. Halff, San Antonio, Texas. This collection includes personal, ranching, and business items relating to Mayer Halff, his relatives, M. Halff & Brother, and Halff & Levy. A significant portion of the Halff Family Collection is on microfilm at Panhandle-Plains Historical Museum, Canyon, Texas.

BOOKS
Adams, Andy. *The Log of a Cowboy*. Garden City, NY: Doubleday & Co., 1964.

Barnes, Charles Merritt. *Combats and Conquests of Immortal Heroes*. San Antonio: Guessaz & Ferlet Co., 1910.

Beck, M. W.; H. W. Hawker; and L. G. Raagsdale. *Soil Survey of Frio County, Texas*. U.S. Department of Agriculture, Series 1929, No. 25.

Beckes, Michael R.; David S. Dibble; and Martha Doty Freeman. *A Cultural Resource Inventory and Assessment of McGregor Guided Missile Range Part 1: The Cultural Resource Base*. Research Report No. 65, Part 1. Austin: Texas Archeological Survey, The University of Texas at Austin, 1977.

Beverly, Bob. *Hobo of the Rangeland*. Lovington, New Mexico, n. d.

———. *To All Boys of U.S.A.* Odessa: Drill Bit Publishing Co., n. d.

Casey, Clifford B. *Alpine, Texas: Then and Now*. Seagraves, TX: Pioneer, 1981.

———. *Mirages, Mysteries and Reality: Brewster County, Texas, The Big Bend of the Rio Grande*. Hereford, TX: Pioneer, 1972.

Coward, Margaret, ed. *The Gaines County Story: A History of Gaines County, Texas*. Seagraves: Pioneer, 1974.

D. & A. Oppenheimer, Bankers (Unincorporated): Transcribed Interviews with Mr. Dan Oppenheimer. Interviews by Larry Meyer. Austin: Oral Business History Project, Graduate School of Business, 1971.

Dary, David. *Cowboy Culture: A Saga of Five Centuries*. Lawrence, Kansas: University Press of Kansas, 1989.

Dearen, Patrick. *Castle Gap and the Pecos Frontier*. Fort Worth: Texas Christian Univ. Press, 1988.

———. *A Cowboy of the Pecos*. Plano: Republic of Texas, 1997.

———. *Crossing Rio Pecos*. Fort Worth: Texas Christian Univ. Press, 1996.

————. *Portraits of the Pecos Frontier*, rev. ed. Lubbock: Texas Tech Univ. Press, 1999.

Dobie, J. Frank. *A Vaquero of the Brush Country*. New York: Grosset and Dunlap, n. d. Originally published by Southwest Press, 1929.

————. *The Longhorns*. Boston: Little, Brown & Co., 1941.

Douglas, C. L. *Famous Texas Feuds*. Dallas: The Turner Co., 1936.

Eagleton, N. Ethie. *On the Last Frontier: A History of Upton County, Texas*. El Paso: Texas Western Press, 1971.

Everett, Donald E. *San Antonio: The Flavor of Its Past, 1845-1898*. San Antonio: Trinity Univ. Press, 1975.

Ford, Gus. L., ed. *Texas Cattle Brands*. Dallas: Clyde C. Cockrell Co., 1936.

Francaviglia, Richard V. *From Sail to Steam: Four Centuries of Texas Maritime History, 1500-1900*. Austin: Univ. of Texas Press, 1998.

Gracy, Alice Duggan; Jane Sumner; and Emma Gene Seale Gentry. *Early Texas Birth Records, 1838-1878*. Austin: Privately printed, 1971.

Halff Family, The. Privately printed, January 1967.

Halff, Harry A., comp. *The Halff Family*. Privately printed, December 1969.

Halff, Henry M. *The Halff Lands: 80,000 Acres in West Texas*. Midland: privately printed, n. d.

Halff, M. H. "Hal." *My Memoirs*. Privately printed, n. d.

Hunter, J. Marvin, ed. and comp. *The Trail Drivers of Texas*. Austin: Univ. of Texas Press, 1985.

Jackson, W. H. and S. A. Long. *The Texas Stock Directory or Book of Marks and Brands Vol. 1*. San Antonio: The Herald Office, 1865.

Jewish Texans, The, 2nd ed. San Antonio: Univ. of Texas Institute of Texan Cultures, 1984.

Kenmotsu, Ray D. and John D. Pigott. *A Cultural Resource Inventory and Assessment of McGregor Guided Missile Range, Otero County, New Mexico, Part 3: Botanical & Geological Studies*. Research Report No. 65, Part III. Austin: Texas Archeological Survey, Univ. of Texas at Austin, April 1977.

Lauderdale, R. J. and John M. Doak. *Life On the Range and On the Trail*. San Antonio: The Naylor Co., 1936.

Llewellyn, Betty Halff, with A. C. Greene. *I Can't Forget: The People, Places and Performances in My Life*. Dallas: Walnut Hill, 1984.

Memorial and Genealogical Record of Southwest Texas. Chicago: Goodspeed Brothers, 1894.

Millard, F. S. *A Cowpuncher of the Pecos*. J. Marvin Hunter, n.d.

Morris, John W. and Edwin C. McReynolds. *Historical Atlas of Oklahoma*. Norman: Univ. of Oklahoma Press, 1965.

Murrah, David. *Oil, Taxes, and Cats: The Saga of the DeVitt Family and the Mallet Ranch*. Lubbock: Texas Tech Univ. Press, 1994.

New Encyclopedia of Texas Vol. 1. Dallas: Texas Development Bureau, n. d.

New Handbook of Texas, The. Austin: Texas State Historical Association, 1996, six volumes.

Oppenheimer, Jesse D. *I Remember*. Privately printed, n. d.

Ornish, Natalie. *Pioneer Jewish Texans*. Dallas: Texas Heritage, 1989.

Partlow, Miriam. *Liberty, Liberty County, and the Atascosito District*. Austin: Pemberton Press, 1974.

Pioneer History of Midland County, Texas 1880-1926, The. Dallas: Taylor, 1984.

Rainey, George. *The Cherokee Strip.* Guthrie, Oklahoma: Co-Operative, 1933.

Roads of Texas, The. Fredericksburg, Texas: Shearer, 1988.

Leonard R. Robbins. *Mèlange: A Genealogical Love Story Vol. 1.* Farkleberry, 1999.

Roberts, J. Travis, Jr., comp., ed. *Early Settlement, People and Place Names, Brewster County, Texas Vol. 1 and 2* (A report to Brewster County Historical Commission 20 August 1998). Marathon, Texas: Privately printed, 1998.

Rosenberg, Stuart E. *To Understand Jews.* New York: Pyramid Books, 1970.

San Antonio City Directory, 1877-1878, 1878-1879, 1881-1882, 1883-1884, 1885-1886, 1887-1888, 1889-1890, 1891, 1892-1893, 1895-1896, 1897-1898, 1899-1900, 1901-1902.

Savage, William W., Jr. *The Cherokee Strip Live Stock Association.* Columbia: Univ. of Missouri Press, 1973.

Skelton, Duford W.; Freeman, Martha Doty; Smiley, Nancy K.; Pigott, John D.; Dibble, David S. *A Cultural Resource Inventory and Assessment of Doña Ana Range, New Mexico.* Research Report No. 69. Austin: Texas Archeological Survey, Univ. of Texas at Austin, 1981.

Sonnichsen, C. L. *I'll Die Before I Run: The Story of the Great Feuds of Texas.* New York: Devin-Adair, 1962.

Stauben, Daniel. *Scenes of Jewish Life in Alsace,* Rose Choron, trans. Joseph Simon Pangloss, 1991. (Originally published in *Revue dès deux mondes* 1857-1859.)

Steinfeldt, Cecilia. *San Antonio Was: Seen Through a Magic Lantern.* San Antonio: San Antonio Museum Association, 1978.

Teter, Gertrude M. and Donald L. Teter, comps., eds. *Texas Jewish Burials.* Texas Jewish Historical Society, 1997.

Thompson, Cecilia. *History of Marfa and Presidio County, Texas, Vol. 1, 1535-1900.* Austin: Nortex, 1985.

Ungnade, Herbert E. *Guide to New Mexico Mountains.* Albuquerque: Univ. of New Mexico Press, 1975.

Wedin, Jo Anne P. *The Magnificent Marathon Basin.* Austin: Nortex Press, 1989.

Williams, Clayton. *Texas' Last Frontier: Fort Stockton and the Trans-Pecos, 1861-1895.* College Station: Texas A&M Univ. Press, 1982.

Winegarten, Ruthe and Cathy Schechter. *Deep in the Heart: The Lives and Legends of Texas Jews.* Austin: Eakin, 1990.

Wyoming Stock Growers' Association. *Brand Book for 1885, Fourth Edition.* Cheyenne: Wyoming Stock Growers' Association, n. d.

CENSUS RECORDS, U.S.
Liberty County 1860
Bexar County 1870

COLLECTIONS
Freeman, Martha Doty Papers. Austin, Texas.
Halff, Henry M. Family File. Nancy McKinley, Midland, Texas.
Haley, J. Evetts. Haley Library, Midland.

James, John Herndon Papers, 1812-1938. The Daughters of the Republic of
   Texas Library, San Antonio, Texas.
Williams, Clayton Wheat. Haley Library, Midland.

COUNTY TAX ROLLS, TEXAS
Atoscosa County, 1894-1905
Bee County 1888-1905
Bexar County 1865-1894, 1905
Crane County 1899
Crockett County 1884-1905
Frio County 1892-1905
Gaines County 1897-1905
Glasscock County 1897-1906
Jefferson County 1862
La Salle County 1892-1905
Liberty County 1857-1870
McMullen County 1887-1905
Medina County 1893-1905
Midland County 1897-1905
Pecos County 1888
Presidio County 1877-1900
Tom Green County 1883-1886
Upton County 1893-1905
Uvalde County 1890-1892, 1898-1905

COUNTY BRAND RECORDS, TEXAS
Pecos County 1899-1901

INTERVIEWS
Allen, Vera Dell. Sterling City, Texas, 8 and 22 February 1998 (both by Patrick
   Dearen).
Bell, Lee. El Paso, Texas, 21 November 1952 (by J. Evetts Haley). Nita Stewart
   Haley Memorial Library, Midland, Texas.
Beverly, Bob. 22 June 1937, 24 March and 13 September 1945, 20 January
   1946, and 23 and 27 June 1946 (by J. Evetts Haley). Haley Library,
   Midland.
Beverly, Walter. 4 March 1998 (by Patrick Dearen).
Boykin, Sid J. 23 June 1927 (by J. Evetts Haley). Haley Library, Midland.
Campbell, Hugh. 30 September 1938 (by Annie McAulay). Federal Writers'
   Project, 1936-1940.
Carr, W. J. D. "Bill." Sherwood, Texas, 12 July 1947 (by J. Evetts Haley). Haley
   Library, Midland.
Chaney, Fred. Garden City, Texas, July 1972 (by Paul Patterson). Southwest
   Collection, Texas Tech University, Lubbock, Texas.
Cauble, Richard. Midland, 20 March 1998 (by Patrick Dearen).
Cochran, Walter. Midland, Texas, date not known (by J. Evetts Haley). Haley
   Library, Midland.
Curry, Lyle. 21 October 1975 (by J. Evetts Haley). Haley Library, Midland.

Evans, W. L. San Antonio, Texas, 9 June 1933 (by J. Evetts Haley). Haley Library, Midland, Texas.

Freeman, Ernestine. 6 May 1998 (by Patrick Dearen).

Graham, E. V. Odessa, Texas, date not known (by J. Evetts Haley), Haley Library, Midland, Texas.

Halff, Albert. 6 May 1998 (by Patrick Dearen).

Halff, M. H. and Mrs. M. H. Halff. Circa December 1975 (by J. Conrad Dunagan). Special Collections, University of Texas or the Permian Basin, Odessa, Texas.

Harral, Ed. Roswell, New Mexico, 13 June 1939 (by J. Evetts Haley). Haley Library, Midland.

Harrington, Elijah Deaver. Pantano, Arizona, 15 December 1939 (by J. Evetts Haley). Haley Library, Midland.

Holmsley, W. H. Midland, 17 October 1926 (by J. Evetts Haley). Haley Library, Midland.

Hudspeth, Roy. San Angelo, Texas, 30 June 1946 (by J. Evetts Haley). Haley Library, Midland.

Johnson, J. A. 27 March 1946 (by J. Evetts Haley). Haley Library, Midland.

Huling, Thomas B. Lampasas, Texas, 28 February 1933 (by J. Evetts Haley). Haley Library, Midland.

Lee, Young. Midland, 3 June 1959 (by J. Evetts Haley). Haley Library, Midland.

Liles, M. L. "Mike." Kenna, New Mexico, 4 August 1937 (by J. Evetts Haley). Haley Library, Midland.

Lockwood, A. L. Tahoka, Texas, date not known (by J. Evetts Haley). Haley Library, Midland.

Meyers, J. S. Mineral Wells, Texas, 22 January 1932 (by J. Evetts Haley). Haley Library, Midland.

McGonagill, Charles. 3 September 1947 (by J. Evetts Haley). Haley Library, Midland.

McNairy, D. H. "Bob." Mineral Wells, Texas, 22 January 1932 (by J. Evetts Haley). Haley Library, Midland, Texas.

McWhorter, Ralph. Lubbock, Texas, 29 November 1979 (by David Murrah and Ann Burkholder). Southwest Collection, Texas Tech University, Lubbock.

Midkiff, Hunter. Midland, 28 April 1998 (by Patrick Dearen).

Midkiff, Melba. Midland, 28 April 1998 (by Patrick Dearen).

Montgomery, Tom. Fort Worth, Texas 2 February 1936 (by Brockman Horne). Haley Library, Midland.

Moore, D. W. Temple, Texas, 2 February 1932 (by J. Evetts Haley). Haley Library, Midland.

Nevill, V. G. Pecos, Texas, 9 July 1968 (by Paul Patterson). Southwest Collection, Lubbock.

Newland, Cliff. Upton County, Texas, 23 August 1964 (by Elmer Kelton). Copy in Patrick Dearen's possession.

Newland, Cliff. Crane, Texas, 15 August 1971 (by Paul Patterson). Southwest Collection, Lubbock.

Nichols, John. Lampasas, Texas, 15 May 1927 (by J. Evetts Haley). Haley Library, Midland.

Nix, Mrs. Will. Rankin, Texas, 30 May 1968 (by Paul Patterson). Southwest Collection, Lubbock.

Oppenheimer, Jesse. San Antonio, Texas, 4 May 1998 (by Patrick Dearen).

Patterson, Paul. Crane, Texas, 27 October 1998; Centralia Draw, Texas, 30 October 1998; and 2 November 1998, Upton County (all by Patrick Dearen).

Roberts, James Travis. 14 April 1886 (by Virgina Wulfkuhle). Archives of the Big Bend, Sul Ross State University, Alpine, Texas.

Roberts, James Travis. 26 July 1986 (by Jim Cullen). Archives of the Big Bend, Sul Ross State University, Alpine, Texas.

Roberts, Travis Jr. 6 May 1998 (by Patrick Dearen).

Steele, A. L. Houston, Texas, 2 November 1932 (by J. Evetts Haley). Haley Library, Midland.

Waddell, R. T. Odessa, 15 December 1962 (by J. Evetts Haley). Haley Library, Midland.

Walker, Walter. 5 August 1937 (by J. Evetts Haley). Haley Library, Midland.

Watson, C. C. Midland, 6 January 1937 (by J. Evetts Haley). Haley Library, Midland.

White, J. Phelps. 2 March 1933 (by J. Evetts Haley). Haley Library, Midland.

Wilson, Jim B. Alpine, Texas, 1 January 1928 (by J. Evetts Haley). Haley Library, Midland.

Windham, A. T. Pecos, Texas, 10 January 1927 (by J. Evetts Haley). Haley Library, Midland.

Wolcott, Henry. Midland, 17 October 1926 (by J. Evetts Haley). Haley Library, Midland.

JOURNALS AND MAGAZINES

Beverly, Bob. "Road Lizzard." *The Cattleman* (October 1946): 109-110.

Crimmins, M. L. "Camp Peña Colorado, Texas." *West Texas Historical and Scientific Society* No. 6 (1 December 1935): 8-22.

Fenley, Florence. "Asa Jones, Early Day Cowman." *The Cattleman* (June 1962).

Fletcher, Henry T. "Violent Early Days of Big Bend Section Recalled." *Frontier Times* (May 1934): 355-357.

Haley, J. Evetts. "Cowboy Sheriff." *The Shamrock* (Summer 1963): 3-6.

Holt, R. D. "Pioneer Cowmen of Brewster County and the Big Bend Area." *The Cattleman* Vol. 29, No. 1 (June 1942): 12-29.

Mahnken, Norbert R. "Ogallala—Nebraska's Cowboy Capital." *Nebraska History, a Quarterly Magazine* Vol. 28 No. 2 (April-June 1947): 85-109.

Modisett, Bill. "Breakup of Quien Sabe promoted settlement of region south of Midland." *Midland Reporter-Telegram* (30 March 1997).

Murchison, Ivan as told to Neighbours, K. F. "Ranching on the Pecos at the Turn of the Twentieth Century." *West Texas Historical Association Year Book* Vol. 53 (1977): 127-136.

Preston, Thomas E. "The Quien Sabe Ranch." *True West* (August 1982): 36-39.

Ridley, Joan. "Get Along, Little Dogies." Undated tear sheet from *Junior Historian*: 17.

Riemenschneider, Larry and Solveig A. Turpin. "Grierson's Spring Military Outpost, 1878-1882." *The Journal of Big Bend Studies* Vol. 10 (1998): 89-123.

Skaggs, Jimmy M. "A Study in Business Failure: Frank Collinson in the Big Bend." *Panhandle-Plains Historical Review* Vol. 43 (1970): 9-19.

White, Grace Miller. "The Activities of M. Halff & Brother." *Frontier Times* Vol. 19 No. 4 (January 1942): 169-175.

LETTERS, MANUSCRIPTS

Arno, F. J. "An Interesting Trip with Meyer Halff in 1884." Clayton Wheat Williams Collection, Nita Stewart Haley Memorial Library, Midland, Texas.

Collinson, Frank to S. R. Coggin. 9 August 1892. Coggin Brothers & Association Records, Southwest Collection, Texas Tech University, Lubbock, Texas.

Collinson, Frank. "My First Experience in Texas." Panhandle-Plains Historical Museum, Canyon, Texas.

Day, Darla. "The Life and Untimely Death of Upland, Texas." Copy in Patrick Dearen's possession.

Duncan, William B. Diary, Sam Houston Regional Library and Research Center, Liberty, Texas.

Fenley, Florence. "Some Horses He Won't Forget; Likewise a Certain Camp Cook: Asa Jones of Brewster County Tells About Them." Special Collections, University of Texas of the Permian Basin, Odessa, Texas. From *Sheep and Goat Raiser* (November 1965).

Guffee, Eddie J. "Camp Peña Colorado, Texas: 1879-1893." Thesis, Archives of the Big Bend, Sul Ross State University, Alpine, Texas.

Haley, J. Evetts. "A Survey of Texas Cattle Drives to the North, 1866-1895." Thesis, June 1926. Haley Library, Midland, Texas.

Kallison, Frances R. "100 Years of Jewry in San Antonio." Thesis, Trinity University, 18 August 1977.

Levy, Helen Blanton and Claudine Murphy. "A Halff Genealogy." Texas Jewish Historical Society Records, Center for American History, The University of Texas at Austin, Austin, Texas.

Love, Jim, as told to Bob Beverly. "Early Day Experience of Jim Love." Bob Beverly file, Haley Library, Midland.

Phillips, Mrs. H. N. "A History of Glasscock County, Texas." Copy in Patrick Dearen's possession.

Preston, Sam to Collie, M. W., Midland, 2 March 1955. Copy in Patrick Dearen's possession.

Starr, E. E. to C. J. Harris, Principal Chief, Cherokee Nation, 20 July 1890. Oklahoma Historical Society, Oklahoma City, Oklahoma.

White, John H. "Cap." "The Activities of M. Halff & Brother," 1942. A paper compiled for G. A. C. Halff. Copy in Patrick Dearen's possession.

MILITARY RECORDS

Post Returns, Camp Peña Colorado, Texas, National Archives, Washington, D.C.

Dandy, G. B., deputy quartermaster general, Headquarters Department of Texas, San Antonio, Texas, to M. Halff and Bro., 1 December 1892, Record Group 393, National Archives.

Elting, O. to assistant adjutant general, Department of Texas, 14 February 1890, Camp Peña Colorado, letters sent, RG 393, National Archives.

Elting, Oscar to adjutant general, 7 May 1890. Letters sent, Camp Peña Colorado, RG 393, National Archives.

Halff, M. & Brother to Lieutenant George H. Morgan, 21 May 1889. Letters received, Camp Peña Colorado, RG 393, National Archives.

Halff, M. & Bro. to Lieutenant George H. Morgan, 22 May 1889. Letters received, Camp Peña Colorado, RG 393, National Archives.

Morgan, Lieutenant George H. to M. Halff & Bro., 21 May 1889. Letters sent, Camp Peña Colorado, RG 393, National Archives.

Morgan, Lieutenant George H.to M. Halff & Bro., 23 May 1889. Letters sent, Camp Peña Colorado, RG 393, National Archives.

Morgan, Lieutenant George H. to assistant adjutant general, Department of Texas, 26 July 1889. Letters sent, Camp Peña Colorado, RG 393, National Archives.

Morgan, Lieutenant George H. to F. H. Oswald, 18 October 1889. Letters sent, Camp Peña Colorado, RG 393, National Archives.

Hess, J. J. to Captain E. G. Fechet, 20 March 1888. Letters received, Camp Peña Colorado, RG 393, National Archives.

NEWSPAPERS

Abilene Chronicle, 1870-1871.

The Caldwell Commercial, 1880.

The Caldwell Post, 1881-1882.

Dallas Morning News, 1887.

Dodge City Times, 1883-1884.

Ford County Globe, 1878-1884.

Freie Presse für Texas, 1870-1884.

Globe Live Stock Journal, 1884-1885.

Liberty Gazette, 1856-1860.

Midland Live Stock Reporter, 10 June 1905.

San Antonio Light (sometimes styled *The Light, The Evening Light*, or the *San Antonio Daily Light*), 1882-1905.

The Pearsall Leader, 29 September 1955.

The Presidio County News (Fort Davis, Texas), 31 May 1884.

Rio Grande Republican, 1885-1893.

Ross's Paper, 1 March 1872.

San Angelo Standard, 1884-1902.

San Angelo Standard-Times, 16 May 1954, 15 September 1957.

San Antonio Express (sometimes styled *Daily Express*), 1877-1905.

The Sidney Telegraph, 1878-1883.

Southern Kansas Advance 27 May 1874.

Texas Live Stock Journal (sometimes styled *Texas Stock and Farm Journal* and *Texas Stockman-Journal*), 1882-1905.

Waco Daily Examiner, 13 July 1877.

Index

163

www.ingramcontent.com/pod-product-compliance
Lightning Source LLC
Chambersburg PA
CBHW052126270326
41930CB00012B/2769